DEMYSTIFYING WATERPROOFING

SANDEEP CHAUDHRY

INDIA • SINGAPORE • MALAYSIA

ISBN 979-8-89363-383-2

Table of Contents

Preface

Having been a Waterproofer since 1999, I have come across many Architects, Home Owners, Contractors, Builders, Waterproofers and waterproofing system manufacturers. While building is being planned and designed, waterproofing is not considered as a specialised item in the BOQ. It is generally kept under the purview of the Civil Contractor. Specialised Teams are involved for Wood Work, Tiling, Stone fixing, metal fabrication, Plumbing, Electrical, False Ceiling, Painting but no separate specialist is considered for Waterproofing. However during the last few years Architects and home owners have started taking this seriously and have started looking for Professional Waterproofers. It has been noted that most of the times the Cost of waterproofing is less the 1% of the total building cost.

Foreword

Though I have been in the architecture profession for over 50 years, certain aspects of building construction, particularly waterproofing, have always been a sensitive issue. The industry, including architects, engineers, builders, and contractors, often lacks a thorough understanding of the waterproofing process. This sensitivity extends from daily waterproofing in basement tank formations to external waterproofing of complete RCC walls, and even internal waterproofing of structures, resulting in major and minor loopholes. Consequently, many basement RCC tanks experience leaks in buildings, requiring regular maintenance on a yearly basis.

I am thrilled and proud that an expert like Shri Sandeep Chaudhry, who is a pioneer and innovator in waterproofing solutions, has taken the lead in publishing a book titled "Demystifying Waterproofing". This book not only delves into the technical aspects but also emphasizes the importance of proper maintenance and quality assurance in new projects as well as existing structures. It addresses the damages caused by various materials like polymers, bituminous compounds, and crystalline mixtures due to leakage.

This publication is a significant contribution, especially to civil engineering, architecture students, and the building industry as a whole. I congratulate Sandeep Chaudhry for his commendable effort in publishing this treasure trove of knowledge, which will serve as a vital source of information for creating healthier and more resilient buildings for tomorrow.

Prof. Charanjit S. Shah
(Founder Principal Creative Group LLP)
(Chairman Smart Habitat Foundation)
(Chairman AKASA Design Centre)

Introduction

Since constructions has started in the world, human being has been striving to protect their homes from the elements of nature such as Water, Air, Heat & Cold and Dust. There has always been a fear that these elements will damage homes. Different geographies have different elements which are of concern for these homes. Despite so much development in technology and advancement of science, people still are not open to use new technologies and products.

Among all the elements, water is the most destructive. While water gives life, it also gives misery. Water and moisture damages more structures than any other calamity in the world. The effects of water damage last longer than any other.

Modern waterproofing systems and technologies help in protecting and preserving a buildings structural integrity throughout its life cycle. A good Waterproofing requires proper analysis of the structural requirement and choosing right product and doing right application. Waterproofing design depends upon various geographical and physical conditions. The effects of natural elements based on geographical and weather conditions of a particular area should be considered while designing a waterproofing system.

There are lots of Myths and misconceptions about waterproofing. Most of these are due to manufacturers not offloading the information with the actual user, home builder, architects, and other stake holders. The Waterproofing being a highly unorganized industry in India, the technology and information has not, yet percolated down to the last person at the site who is the actual technician applying the product. Some advertisements in Print and other media also show cases that a few products are the only waterproofing solution and professional application, and product selection is not required. This has caused lot of product and application failures on the sites, which has resulted in stake holders either losing confidence in new products and technologies or taking wrong decisions about using them. In the chapters which will follow, we will try to Demystify Waterproofing and make it easy to understand and take calculated and conscious decisions.

Waterproofing depends upon not only the use of the products and application. It is also important to integrate various other structural detailing. Basic waterproofing principals such as slope maintenance also plays major role in protecting structures from seepages. Other services and trades which go into building a structure should also be integrated while designing and doing waterproofing.

Before we move ahead, lets know some of the common terms used in waterproofing specifications:

Below Grade Waterproofing: Commonly known as Basement Waterproofing, it is the waterproofing of the areas which go under the soil. The material used prevent and withstand Hydro Static pressure to let water enter the sub structure. These systems should also be able to withstand the chemicals and impurities commonly found in the ground water. Below grade structure also include Under Ground water storage tanks and swimming pools.

Damp Proofing: Damp Proofing is a moisture control system used to prevent water moisture getting into the building. This is one of the most frequent issues in buildings.

Seepage: Seepage is generally thru small holes or porosity in the structure. Once again, a major cause of worries.

Slope Maintenance: is creating slopes on the roof surface to help drain water fast, without causing any water accumulation. It is said that a good slope can solve 50% water proofing problems.

Diversions: Diversions redirect water from various building elements and divert it away from the structure. Drain Cells, Drainage Boards, Spouts, Drainpipes are some of these.

Negative Side Waterproofing: When waterproofing is applied from the reverse side of possible water ingress

Positive Side Waterproofing: When waterproofing is applied from the side from where water ingress is possible.

Membrane: Generally, Pre-Formed membranes are called Membranes, such as Tar Felts, APP, EPDM, HDPE, TPO etc. These are factory manufactured products which come in different sizes 1.0 mtr to 3.0 mtr wide and 20-30 mtrs length. Thickness too may vary from 1.0 mm to 4.0 mm

Liquid Applied Membranes: are membranes which are applied as a coating. The coatings look like a single membrane. Products which are used for LAM are Polyurethane, Pitch Modified Polyurethane, Acrylic Coatings etc.

Cementitious Coatings: Coatings which are a blend of Cement and Polymers. The Polymer part could be Latex or Acrylic.

Two Component Cementitious Coatings: Coatings which are Factory blended set of Powder and Polymer. The powder is a blend of Cement, Fine Aggregates, Dry Polymers, and specialty Fibers. The blend depends on manufacturer to manufacturer. A fixed mixing ratio of Powder and Polymer is given and is mixed at site accordingly.

Crystalline Coating: is a technology which is used primarily for an additional protection of underground concrete structure using specialized cementitious crystalline mixture.

Grouts & Joint Fillers: There are different types of grouts used in construction. The grouts are generally a mixture of Cement and Sand. These are mixed at site with water and Polymer, and are used to fill different types of Gaps, Joints, Cracks etc., These are also used to give additional strength to some types of

Admixtures: These are different types of Chemical admixtures to provide exceptionally durable & predictable concrete. These help in improving the quality and strength of concrete.

Bonding Agents: These are Epoxy or Polymer based bonding agents in varied viscosities to ensure proper bonding.

Concrete Repair: There is a range of pre-mix concrete repair products for patch repairs, form & pour and resurfacing solutions for vertical, overhead, horizontal, and underwater concrete repairs.

Injection Grouting: Injection grouting is when the structure – Concrete or brick has some voids or cracks which need to be injected with suitable cementitious, polyurethane, epoxy, and other chemicals to fill the voids and cracks.

Surface Preparation: is the most important and crucial component in waterproofing. A properly prepared surface enhances the performance of any waterproofing system. The surface should be cleaned properly using mechanical grinding to remove mortar buildup from the surface. The surface should be cleaned so that sharp edges, latience and construction residue is removed. The cracks, pin holes and perforations should be filled with acrylic modified repair mortars. Any powdery substance should be removed using a vacuum cleaned. The surface should be washed using water pressure jets. The surface should be kept in SSD (Surface saturated Dry) condition.

Elongation: refers to the ability of the coating to stretch or extend without breaking when subject to stress or deformation. A higher elongation value indicates greater flexibility and durability of the coating.

Tensile Strength: refers to the maximum amount of tensile stress that a coating can withstand before it breaks or fails. Its an important mechanical property of coatings as it indicates their ability to withstand external forces and maintain structural integrity under tension. A higher tensile strength value indicates greater durability.

Tear Strength: refers to the resistance of a material to tearing or rupture when subject to a tensile force. A higher tear strength indicates greater resistance to tearing and better overall durability of the material. In waterproofing coatings, tear strength is essential for ensuring long term performance and resistance to damage from external factors such as impact, abrasion, or mechanical stress.

Compressive Strength: refers to the maximum amount of stress that a material can withstand before it fails or collapses under compression.

Breathable: refers to the ability of the coating to allow water vapour or air to pass through while still providing a barrier against water.

Crack Bridging: refers to the ability of a material or coating to span overand maintain its integrity across existing cracks or fissures in a substrate.

Solid Contents: refers to the percentage of non-volatile components present in chemicals formulation. These non-volatile components typically include resin, polymers, fillers and other active ingredients that contribute to the properties and performance of the waterproofing products.

Fully Bonded System: refers to waterproofing membranes or coatings which are fully adhered or bonded to the substrate.

Chapter 1

Introduction to Waterproofing

1.1 Understanding the Importance of Waterproofing

India's vast and diverse landscape encompasses a wide range of climatic conditions, from the arid deserts of Rajasthan to the snow-capped peaks of the Himalayas, presenting unique challenges for waterproofing construction projects. In this chapter, we delve into the importance of waterproofing in the Indian context, where the weather can vary significantly from region to region, necessitating tailored solutions to protect structures against moisture intrusion, water damage, and structural deterioration.

1. Varied Climatic Conditions:

India experiences diverse climatic zones, each characterized by distinct weather patterns and environmental factors:

- Tropical Regions: Coastal areas and regions in the south experience hot and humid climates, with heavy rainfall during the monsoon season.
- Arid and Semi-Arid Zones: North western and central regions of India are prone to arid and semi-arid climates, with low rainfall and high temperatures.
- Mountainous Terrain: The Himalayan regions in the north witness extreme cold temperatures, snowfall, and freeze-thaw cycles, posing unique challenges for waterproofing.

2. Differential Waterproofing Needs:

The varied climatic conditions across India necessitate differential waterproofing solutions tailored to the specific requirements of each region:

- Coastal Areas: Waterproofing solutions for coastal regions must prioritize resistance to saltwater intrusion, UV radiation, and high humidity levels to prevent corrosion, degradation, and mold growth.
- Desert Regions: In arid and semi-arid zones, waterproofing products should focus on thermal insulation, UV resistance, and crack-bridging capabilities to withstand temperature extremes and prevent water seepage.
- Himalayan Regions: Waterproofing systems in mountainous terrain must be engineered to withstand sub-zero temperatures, heavy snow loads, and freeze-thaw cycles, necessitating robust materials with superior flexibility, durability, and resistance to ice formation.

3. Tailored Waterproofing Solutions:

Architects and construction professionals must assess the specific climatic conditions and environmental factors prevalent in each region to select appropriate waterproofing chemicals and materials:

- Liquid-Applied Membranes: Flexible and seamless liquid-applied membranes are ideal for coastal areas, providing superior waterproofing and protection against saltwater intrusion and UV exposure.
- Polyurethane Coatings: Polyurethane coatings offer excellent adhesion, durability, and resistance to temperature fluctuations, making them suitable for use in desert regions where thermal insulation is crucial.
- Cold-Applied Membranes: Cold-applied membranes are specifically designed for use in cold climates, offering superior

flexibility and crack-bridging properties to withstand the challenges of snowfall and freeze-thaw cycles.

4. Professional Expertise:

Collaboration with experienced waterproofing contractors is essential to ensure the effective implementation of tailored solutions suited to the diverse climate of India. Professional waterproofers possess the technical knowledge, skills, and industry certifications necessary to execute waterproofing projects with precision, quality, and compliance.

In conclusion, the importance of waterproofing in the Indian context cannot be overstated, given the diverse climatic conditions prevalent across the country. By understanding the unique challenges posed by varying weather patterns and environmental factors, architects and construction professionals can select and implement customized waterproofing solutions that protect structures against moisture intrusion, water damage, and structural deterioration, thereby ensuring the longevity and integrity of buildings in the face of India's diverse climate.

1.2 Challenges in the Waterproofing Industry: Navigating Material Selection and Application

The waterproofing industry is fraught with challenges for architects, clients, and waterproofing professionals alike, stemming from confusion surrounding material selection, application methods, and the proliferation of do-it-yourself (DIY) products marketed to unsuspecting consumers. In this chapter, we delve into the complexities and pitfalls faced by stakeholders in the waterproofing process and explore the impact of misinformation, unqualified contractors, and substandard products on project outcomes and long-term durability.

1. Confusion in Material Selection:

Architects and clients often grapple with the daunting task of selecting the right waterproofing materials amidst a sea of options, each promising superior performance and longevity. Manufacturers' advertisements for DIY products further compound the confusion, leading to misconceptions about the suitability of certain materials for specific applications.

2. DIY Misinformation:

The rise of DIY waterproofing products marketed directly to consumers has blurred the lines between professional-grade materials and consumer-grade solutions. While DIY products may seem convenient and cost-effective, they often lack the durability, quality, and expertise required for long-lasting waterproofing solutions. Architects and clients must be wary of the limitations of DIY products and prioritize collaboration with experienced waterproofing professionals for optimal results.

3. Challenges for Waterproofing Professionals:

Experienced waterproofing professionals face numerous challenges in their line of work, including:

Competition from Unqualified Contractors:

Waterproofing professionals often find themselves competing with civil contractors, plumbers, and painters who lack the specialized knowledge and training necessary for effective waterproofing. These unqualified contractors may offer substandard products and services at lower prices, undercutting the expertise and quality assurance provided by professional water proofers.

Misapplication of Products:

Improper application techniques and inadequate surface preparation can compromise the effectiveness of waterproofing materials, leading to premature failure, water ingress, and costly remediation efforts. Waterproofing professionals must educate clients about the importance of proper installation and adherence to manufacturer specifications to ensure the longevity and performance of waterproofing systems.

Chapter 2

Principles of Waterproofing

2.1 Moisture Sources and Movement

In the first section of the chapter "Principles of Waterproofing" in the book, the discussion would focus on the various sources of moisture ingress into buildings and how water moves through different building materials. This section would likely cover topics such as rainwater penetration, groundwater seepage, condensation, and capillary action.

To track the movement of water and check for moisture, different methods and technologies can be employed:

1. Visual Inspection: Regular visual inspections can help identify signs of water damage such as stains, discoloration, peeling paint, or mold growth.
2. Moisture Meters: Moisture meters are handheld devices that measure the moisture content of building materials such as wood, drywall, and concrete. They can help detect moisture levels and pinpoint areas of concern.
3. Infrared Thermography: This technique uses infrared cameras to detect temperature variations in building materials, which can indicate areas of moisture infiltration.
4. Probe Tests: Probe tests involve inserting probes into building materials to measure moisture content at different depths.

These tests can provide detailed information about the extent of moisture intrusion.

5. Dye Tests: Dye tests involve injecting coloured dyes into suspected areas of water ingress and then visually inspecting for any signs of dye penetration, which indicates the path of water movement.

6. Electronic Leak Detection: Electronic leak detection systems use sensors and conductive materials to detect leaks in waterproofing membranes or plumbing systems.

7. Building Envelope Analysis: Comprehensive analysis of the building envelope, including roof, walls, windows, and foundations, can help identify potential points of water entry and assess the overall waterproofing effectiveness.

By employing these methods and techniques, building owners and professionals can effectively track the movement of water and identify areas susceptible to moisture intrusion, allowing for timely remediation and waterproofing measures.

2.2 Key Concepts in Waterproofing

In the second section of the chapter "Principles of Waterproofing" in the book, the focus is on the key concepts of waterproofing, including the selection of appropriate waterproofing products and methods of application. This section covers the following aspects:

1. Selection of Waterproofing Products: Discussing the various types of waterproofing materials available, such as liquid-applied membranes, sheet membranes, cementitious coatings, and polyurethane sealants. Factors to consider when selecting a waterproofing product include the type of substrate, environmental conditions, durability requirements, and budget constraints.

2. Surface Preparation: Emphasizing the importance of proper surface preparation before applying waterproofing materials.

This may involve cleaning the surface to remove dirt, debris, and contaminants, repairing cracks and defects, and ensuring a smooth and uniform substrate.

3. Application Methods: Explaining the different methods of applying waterproofing materials, including brush application, roller application, spray application, and trowel application. Each method has its advantages and is suitable for specific types of waterproofing products and substrates.

4. How to Apply the Material: Providing step-by-step instructions on how to properly apply the selected waterproofing product, including mixing ratios (if applicable), application temperature and conditions, curing times, and proper techniques to ensure uniform coverage and adhesion.

5. Important Considerations: Highlighting key considerations to keep in mind during the waterproofing process, such as:

   - Weather conditions: Avoiding application during rainy or excessively humid weather.
   - Safety precautions: Following recommended safety guidelines, such as wearing protective clothing, gloves, and goggles, and ensuring proper ventilation in confined spaces.
   - Compatibility: Ensuring compatibility between the waterproofing product and the substrate material to prevent adhesion issues or chemical reactions.
   - Quality control: Implementing quality control measures to verify proper application and adherence to manufacturer specifications.

By understanding these key concepts and following best practices for product selection, surface preparation, and application methods, individuals involved in waterproofing projects can achieve effective and long-lasting waterproofing solutions for buildings and structures.

Chapter 3

Types of Waterproofing Materials

In the realm of construction and building maintenance, ensuring effective waterproofing is paramount to safeguarding structures against moisture infiltration and its detrimental consequences. Chapter 3 delves into the diverse array of waterproofing materials available, each tailored to specific applications, substrates, and environmental conditions. Understanding the characteristics and benefits of these materials is essential for architects, engineers, contractors, and building owners seeking reliable solutions for their waterproofing needs.

This chapter serves as a comprehensive guide to the various types of waterproofing materials, encompassing liquid-applied membranes, sheet membranes, cementitious coatings, polyurethane sealants, and more. Each material offers unique properties and advantages, allowing professionals to select the most suitable option based on factors such as substrate composition, exposure to weather elements, and project requirements.

From flexible and elastomeric membranes capable of accommodating structural movements to rigid and durable coatings ideal for heavy-duty applications, the range of waterproofing materials caters to diverse construction scenarios. Furthermore, advancements in technology continue to drive innovation in waterproofing, yielding products with enhanced performance, durability, and environmental sustainability.

Throughout this chapter, readers will gain insight into the characteristics, applications, installation methods, and maintenance considerations associated with various types of waterproofing materials. By equipping themselves with this knowledge, stakeholders can make informed decisions to effectively protect buildings and infrastructure from water damage, ensuring their longevity and resilience in the face of nature's elements.

3.1 Liquid Applied Membranes

Liquid applied membranes represent a versatile and widely used category of waterproofing materials, offering a flexible and seamless barrier against moisture ingress. This section explores various types of liquid applied membranes, including acrylic, elastomeric modified rubber, polyurethane, and polyurea, highlighting their unique properties, benefits, as well as their respective pros and cons.

Liquid Applied systems can be both Solvent and Water based materials. These may contain base of Polyurethane, Rubbers, Bitumen, Acrylics and /or a combination of these. These are in liquid form and after curing form a seamless sheet.

Liquid Applied systems are positive side waterproofing application and require a protective layer. These are used primarily because of ease of application, seamless curing, and adaptability to detailing. These systems are popular for basement, podiums, planters, and garden application. The systems used for basement waterproofing are not UV Stabilized and cannot withstand foot traffic. These systems typically have an elongation of over 400% with recognized testing as per ASTM C 836. This enables these systems to bridge cracks <2mm. The Liquid Systems have self- flashing capabilities which enables these to be applied seamless at substrate protrusions, changes in planes and floor wall junctions. Some of the Liquid Applied Systems are.

1. **Acrylic Membranes**: These are Single Component water based systems. The substrate required could have some moisture. A SSD (Saturated Surface Dry) condition is most suitable as it helps acrylics to penetrate the surface. These coatings have excellent adhesion, elongation and tensile strength

- Properties: Acrylic membranes are typically water-based and form a breathable, protective coating upon application. They are known for their excellent UV resistance and flexibility.
- Benefits: Acrylic membranes offer ease of application, quick drying times, and compatibility with a wide range of substrates. They are also cost-effective and environmentally friendly.
- Advantages: Good resistance to weathering, excellent adhesion, and compatibility with other coatings.
- Disadvantages: Limited elongation and lower durability compared to elastomeric membranes.

2. **Elastomeric Modified Rubber Membranes**: Rubber derivative systems are compounds of Butyl, neoprene's or Hypalon's. These are Solvent based and have excellent elastomeric properties. These have elongation between 300% to 600%. These are resistant to environmental chemicals which are generally found in soil. Some manufacturers have also started manufacturing Water based Rubber derivative products.

- Properties: Elastomeric modified rubber membranes combine rubber polymers with other additives to enhance flexibility, elongation, and durability. They can accommodate structural movements and bridge cracks effectively.
- Benefits: Exceptional elongation properties, superior flexibility, and resistance to cracking and tearing. Suitable for both vertical and horizontal applications.
- Advantages: Excellent waterproofing performance, long-lasting durability, and high tensile strength.

- Disadvantages: Higher material cost compared to acrylic membranes, longer curing times, and sensitivity to temperature fluctuations during application.

3. **Polyurethane Membranes**: These are available in Two component and Single Component both as Solvent and Water Based systems. The substrate requires to be completely dry. Though some manufacturers have started manufacturing system for wet surface application too. In case of systems which need dry surface, moisture in the surface may cause blisters in the film. Polyurethanes have the highest elongation in range of 350%-600% and some may be more then this and the best crack bridging capabilities. These are Chemical and Alkali resistant.

- Properties: Polyurethane membranes are solvent-based or moisture-cured and form a tough, elastic membrane upon curing. They exhibit excellent adhesion and chemical resistance.
- Benefits: Superior mechanical properties, including high tensile strength and elongation. Suitable for applications requiring resistance to abrasion, chemicals, and UV exposure.
- Advantages: Seamless and monolithic application, fast curing times, and versatility in application methods.
- Disadvantages: Vulnerable to moisture during curing, requiring proper surface preparation and application in dry conditions.

4. **Polyurea Membranes:**

- Properties: Polyurea membranes are fast-curing, spray-applied coatings that form a highly durable, seamless membrane. They exhibit exceptional chemical resistance and are ideal for high-traffic areas.
- Benefits: Rapid curing times, even in low temperatures, and excellent adhesion to various substrates. High tensile strength and resistance to abrasion, chemicals, and extreme weather conditions.

- Advantages: Seamless application, minimal downtime, and excellent performance in harsh environments.
- Disadvantages: Higher material cost compared to other liquid applied membranes, specialized equipment required for application.

5. **Pitch Modified Urethanes**: Bitumen/Pitch/Coal Tar modified urethanes systems reduce the cost of material while still providing similar performance. These have lower elastomeric properties and are not UV Stabilized. These provide durable and seamless waterproofing coatings.

Advantages of Liquid Applied Coatings:

1. Seamless / Monolithic Membrane
2. Shorten Installation Phase
3. Superior Waterproofing
4. Adaptable to detailing
5. Excellent Elastomeric properties
6. Ease of application

Disadvantages of Liquid Applied Coatings:

1. Application thickness controlled at site
2. Some products can not apply on wet surfaces

In conclusion, liquid applied membranes offer an effective solution for waterproofing applications, with each type possessing unique properties and benefits suited to specific project requirements. By understanding the characteristics and considerations associated with acrylic, elastomeric modified rubber, polyurethane, and polyurea membranes, stakeholders can make informed decisions to achieve reliable and durable waterproofing solutions for buildings and infrastructure.

3.2 Pre-formed Membranes

Preformed membranes offer a reliable and efficient solution for waterproofing applications, providing a durable barrier against moisture infiltration. This section explores various types of preformed membranes, including APP (Atactic Polypropylene), PVC (Polyvinyl Chloride), HDPE (High-Density Polyethylene), SBS (Styrene-Butadiene-Styrene), and non-bituminous composite membranes. Each type of membrane exhibits distinct properties, application methodologies, as well as pros and cons, influencing the selection of resin for specific projects.. These are popular choice as these can withstand high level of ground water pressure, aggressive chemicals present in soil or water, dynamic forces due to earthquakes and ground settlements.

The sheet membranes come in fixed and controlled thicknesses, widths, and lengths. These range in thicknesses ranging between 1mm to 5mm and widths ranging between 1.0 mtr to 3.0 mtr.

Application of Pre-Formed Membranes, in the basement requires protection during backfilling, Steel Binding and Concrete Pouring. Some manufacturers, in the recent years have started offering products for below the raft application, which they claim to not need any protection from Steel Binding and Concrete Pouring. Unlike Liquid Applied Membranes, Preformed Membranes have multiple joints and are not seamless. These are also not self-flashing at protrusions and junctions. Same is the case at the termination pr transition into another section of the building envelope. Being too heavy these are limited for application on the horizontal surfaces. Vertical single ply membranes are more difficult to apply then the Liquid Applied membranes. It is difficult to handle and joining the materials. The seams are lapped and sealed for complete waterproofing, particularly in small areas such as planters.

Different types of Pre-Formed membranes are.

- Thermoplastic
- Vulcanized Rubbers
- Polymer Modified Bitumen
- Non Bituminous Composite

THERMOPLASTICS:

Thermoplastic Sheet membranes are available in three types. PVC, HDPE and TPO. These are all highly flexible sheet membrane system and protect any concrete structure. These can be built as single layer compartment systems up to the active control system to meet the highest demand. These can be loosely laid or fully bonded, depending upon product to product. The application process is extremely fast with all these membranes. Extremely high flexibility allows easy installation and completion of details that results in faster membrane installation.

These are available in widths ranging from1.0 mtr-3.0mtr. On horizontal applications, wider roll widths help in minimizing overlap joints. All these systems adhere by solvent based adhesives or heat welding at seems. Some manufacturers have started offering self-sealing seams

All the three derivatives have excellent hydrostatic and chemical resistance to below ground application conditions. PVC membranes are generically brittle materials, but elongation of all this system is acceptable for below ground conditions.

PVC (Polyvinyl Chloride) Membranes:

- Properties: PVC membranes are lightweight, flexible, and resistant to UV radiation, chemicals, and root penetration. They are available in various thicknesses and colours.
- Application Methodology: PVC membranes are typically adhered to the substrate using adhesive or hot-air welding techniques. Proper surface preparation and seam welding are critical for achieving a seamless waterproofing system.

- Benefits: Excellent durability, resistance to punctures, and ease of maintenance.
- Disadvantages: Susceptible to plasticizer migration over time, potentially leading to embrittlement and reduced flexibility.

HDPE (High-Density Polyethylene) Membranes:

- Properties: HDPE membranes are highly resistant to punctures, chemical degradation, and root penetration. They are commonly used in below-grade applications such as foundation walls and tunnels.
- Application Methodology: HDPE membranes are typically loose-laid or mechanically fastened to the substrate, with overlapping seams sealed using heat welding or specialized tapes.
- Advantages: Excellent puncture resistance, flexibility, and longevity.
- Disadvantages: Limited elongation, requiring careful detailing and installation to accommodate structural movements.

VULCANIZED RUBBERS:

Vulcanized Rubbers are available in Butyl, EPDM and Neoprene rubbers. Among these EPDM Sheets are the most popular and accepted with the applicators. These materials are vulcanized to achieve better elasticity and durability. These are available in 1mm-2mm thicknesses. These materials are not breathable and will de bond or blister if negative vapor pressure is present. These are available in widths ranging from 1.0 mtrs- 3.0 mtrs. The seam sealing is with a solvent based adhesive. These cannot be heat welded. For vertical areas these need to be fully bonded using adhesives. For horizontal surfaces these can be laid loosely or fully bonded. Loosely laid applications increase the elasticity capability of the membrane as compared to fully adhered membranes have restricted movement.

POLYMER MODIFIED BITUMEN:

Polymer modified bitumen membranes are manufactured from a blend of Bitumen and selected Polymers. Generally, these come with two Polymer blends. APP (Atactic Polypropylene) and SBS (Styrene Butadiene Styrene)

- **APP (Atactic Polypropylene) Membranes**: APP Membranes is a plastic bitumen blend which is modified using plastic in the form of Atactic Polypropylene. As installers begin to torch apply APP, the plastic begins to melt, where it melts into a liquid wax like substance which acts as an almost free flowing liquid which can be mopped across a surface. APP has high temperature tolerance and is user friendly to the applicators.

The APP membranes have improved thermal properties, better durability, plasticity, resistance to low & high temperatures, and aging. It has high resistance to hydrostatic pressures. These are available with three reinforcements such as Fiberglass, Polyester and Polyester Mat & HMHDPE. These are also available in different top finishes such as Black Top, Minerals, Geo Textile, Mechanically Fixed, Anti Root and Sand Covered. Thickness varies from 1.5 to 5mm.

- Application Methodology: Heat welding or torch application is commonly used to adhere APP membranes to the substrate. Proper surface preparation and seam sealing are essential for ensuring watertight installations.
- Advantages: High tensile strength, flexibility, and resistance to aging and environmental factors.
- Disadvantages: Vulnerable to high temperatures during installation, requiring careful handling to prevent damage.
- **SBS (Styrene-Butadiene-Styrene) Membranes:** SBS Modified Bitumen consists of bitumen with a synthetic rubber modification, blending styrene butadiene styrene. This melts as a sticky substance which increases the hotter it gets. This lack of flowability means it takes less heat to install and allows

a faster installation process. This can also be installed with a cold adhesive when open flames are not allowed.

Because of SBS used in its blend, SBS Bitumen is more flexible compared to APP. Because of flexibility it has more recovery properties, making it capable of withstanding stresses created by wind, temperature fluctuations, Expansions and Contractions.

- Properties: SBS membranes feature rubberized asphalt modified with styrene-butadiene-styrene polymers, offering enhanced flexibility, elongation, and crack-bridging capabilities.
- Application Methodology: SBS membranes are torch-applied or cold-applied using adhesives, with heat welding or seam tape used to seal overlaps and joints.
- Advantages: Superior flexibility, resistance to thermal cycling, and compatibility with irregular substrates.
- Disadvantages: Susceptible to UV degradation unless protected by a surfacing material, such as gravel or pavers.

Benefits of Polymer Modified Bitumen membranes:

1. Longevity
2. Strength
3. Elasticity- Good elongation and recovery
4. Rot proof
5. Excellent puncture resistance

NON-BITUMINIOUS COMPOSITE MEMBRANE: are a non-adhesive non bituminous composite membrane.

- Properties: Non-bituminous composite membranes are manufactured with a core of TPO, or thermoplastics and reinforcement layer of fibres on both sides for enhanced durability and performance.
- Application Methodology: Non-bituminous composite membranes are typically adhered using Type 2 Cementitious Adhesives.

- Advantages: Versatility in material selection, chemicals, and root penetration. Non-toxic, no pollution, environment protecting. Excellent adhesion to most of the surfaces. Tough, impermeable, multi-layered sheet, High tensile strength and good cold flexibility. Thin, flexible, compatible with common cement screeds and resistant to most common alkaline solutions.

- Disadvantages: Has to be covered and cannot be left exposed, requiring skilled installation for optimal performance.

When selecting the resin for each type of preformed membrane, considerations such as project requirements, environmental conditions, substrate compatibility, and budget constraints should be taken into account. By understanding the properties, application methodologies, and pros and cons of each type of preformed membrane, stakeholders can make informed decisions to achieve durable and effective waterproofing solutions for various construction projects.

3.3 Cementitious Coatings

Cementitious coatings offer a versatile and cost-effective solution for waterproofing various substrates, including concrete, masonry, and even metal surfaces. This section explores three types of cementitious coatings: Two Component Cementitious, Two Component Crystalline Coatings, and Cement Polymer Coatings. We delve into why and when to select these coatings, application methods, as well as their pros and cons, including areas where they are not suitable for use.

Cementitious materials are excellent products to be applied on below grade structures, both positive and negative sides. These systems are based on a mix of Portland Cement with or without sand and an active waterproofing agent. There are two types of cementitious systems: Crystalline System, Acrylic Modified Systems. These are applied using brush or trowel to the concrete or masonry surfaces and become integral part of the structure.

In the under-construction buildings, the biggest benefit of using a Cementitious system is that it does not need a dry surface. In fact, some moisture in the structure helps the system and improve its efficacy. It can also be applied if the concrete is not fully cured. These systems can be applied to both the walls and floor

All cementitious systems have similar application method. The performance may differ from product to product depending upon its formulation. The Cementitious systems lack crack bridging or elastomeric properties but have great adhesion to the substrate.

1. **Acrylic Modified Systems:**

Acrylic modified cementitious systems are generally called Two Component Cementitious systems. These are essentially are mix of a factory blended mix of Portland Cement & Sand (may or may not be blended with special fibbers) and Acrylic Emulsions. These acrylic emulsions add waterproofing characteristics to the system. These systems can be applied with brush or trowel. These can also be applied with a reinforcement mesh. The mesh adds some crack bridging capabilities. Many manufacturers are offering premix special fibbers in the powder component. Different manufacturers offer product with Powder: Polymer ratio from 1: 0.2 to 1: 0.5. Higher the Polymer ration, better would be elongation and crack bridging capacity.

Acrylic Modified coating are applied in a thickness of 1.0mm to 2.0 mm. However, in certain application, using a trowel these can be applied up to 4-5mm thickness too. It is highly recommended to apply protection layer. Curing is recommended for 24 hours.

Advantages of Cementitious Systems:

- Positive or Negative Side application
- Remedial/ Repair Application

Disadvantage of Cementitious Systems:

- – No Movement Capability
- – Site Mixing Required
- – Can Not be used in high traffic areas

2. Two component cementitious coatings: are a factory packed pre measured set of Dry Powder and Liquid Polymer. The Powder component is mixture of Cement and graded sand, may or may not have glass fibre. The liquid polymer is a mix to formulated polymer. These are ideal for waterproofing concrete structures, such as basements, retaining walls, and swimming pools. They provide excellent adhesion to concrete substrates and offer protection against water infiltration. These are available in different ratio of Powder and Liquid Polymer. Typically these are available in packs of 15 kg. (10kg.Powder + 5 kg. Liquid Polymer), 27 kg. (23 kg. Powder + 4 kg. Liquid Polymer) and 20kg. (15kg.Powder +5 kg. Liquid Polymer). However, a few manufacturers have launched a high strength system in 20 kg. packs with 10 kg. Powder + 10kg. Liquid Polymer. These have high elongation as well as tensile strength.

- – Application: These coatings are typically applied in two stages: first, a liquid polymer component is mixed with a cement-based powder component to form a slurry, which is then applied to the prepared substrate using a brush, roller, or spray equipment.

These products have a elongation between 25% to 225% and have good tensile strength. The crack briding capacity is up to 2mm. A sturdy product with great adhesion properties.

3. Crystalline Systems are a mixture of Cement and Sand and have a combination of proprietary chemicals. The systems are applied by Trowel, Brush, or spraying. These have, unlike other Cementitious systems, an added advantage that these can be even applied as a dry mix by broadcasting before or after the concrete slab is casted, before the concrete reaches its final set. This is commonly called Dry Shake method. The Crystalline systems forms a crystalline structure within the concrete using water and calcium hydroxide present in the concrete. These crystalline structure block water transmission through the substrate and add additional water repellence.

Though the chemical process begins immediately, but it may take 30-40 days to fully reach maximum repellence. Once fully cured, crystalline systems have been tested to withstand hydrostatic pressure as a high as 400 ft. of water head.

These systems do not need.

- Protection layer
- Seal Hairline cracks, which occur after the application
- Resistant to Chemicals and acids
- Reacts with concrete to form additional protection.

Important to note.

- Crystalline systems need water for the crystal growth. It is important to cure using a spray pump for a minimum of 48 hours.
- The concrete surface should be either wet or uncured for best results.
- Do not leave exposed in interiors or exterior and cover with Drywall or Panelling. Floor surface should be protected with Tiles/Stone/Carpet/Floor Screed.

Two Component Crystalline Coatings:

- Selection: Two component crystalline coatings are suitable for waterproofing below-grade structures, such as foundations, tunnels, and sewage treatment facilities. They react with moisture in the substrate to form insoluble crystals, effectively blocking water ingress.

- Application: These coatings are applied in two stages: first, a liquid activator component is mixed with a cement-based powder component to form a slurry, which is then applied to the prepared substrate using a brush, roller, or spray equipment.
- Advantages: Self-healing properties, long-lasting waterproofing performance, and resistance to hydrostatic pressure. Due to the liquid component, the system also creates a top layer on the surface to repel water.
- Disadvantages: Require proper surface preparation and application to ensure uniform crystalline growth, limited application in areas with continuous exposure to pressurized water.

Single Component Crystalline Coatings:

- Selection: One component crystalline coatings are suitable for waterproofing below-grade structures, such as foundations, tunnels, and sewage treatment facilities. They react with moisture in the substrate to form insoluble crystals, effectively blocking water ingress.
- Application: These coatings are applied in two stages: Cement-based crystalline powder component is mixed with water to form a slurry, which is then applied to the prepared substrate using a brush and roller in two coats.
- Advantages: Self-healing properties, deep penetration, long-lasting waterproofing performance, and resistance to hydrostatic pressure.
- Disadvantages: Require proper surface preparation and application to ensure uniform crystalline growth, limited application in areas with continuous exposure to pressurized water. Require curing They require careful consideration of application requirement and limitations.

Cement Polymer Coatings:

- Selection: Cement polymer coatings are suitable for waterproofing both horizontal and vertical surfaces, including roofs, balconies, and facades. They combine the durability of cement with the flexibility of polymers, offering enhanced crack resistance and adhesion.
- Application: These coatings are typically applied in a single layer using a trowel or spray equipment, with additional layers applied as needed to achieve the desired thickness and coverage.
- Advantages: Excellent adhesion to a variety of substrates, flexibility to accommodate substrate movement, and resistance to UV radiation and weathering.
- Disadvantages: Limited breathability, potential for delamination if not applied properly, and higher material cost compared to traditional cementitious coatings.

Areas Where Cementitious Coatings Are Not Suitable:

- Surfaces with continuous exposure to pressurized water, such as swimming pools and water tanks.
- Surfaces with significant movement or vibration, as cementitious coatings may crack or delaminate over time.
- Surfaces with inadequate substrate preparation or structural integrity, as proper surface preparation is essential for adhesion and performance.

In conclusion, cementitious coatings offer a versatile and durable solution for waterproofing various substrates, with each type catering to specific project requirements and conditions. By understanding their properties, application methods, and limitations, stakeholders can effectively select and utilize cementitious coatings to achieve long-lasting waterproofing solutions for a wide range of construction projects.

3.4 Concrete Admixtures

Concrete Admixtures are used to simplify the protection process of concrete to reduce its permeability. This can make the concrete waterproof itself. A variety of admixtures are available with most of the Construction Chemical manufacturers. But broadly three types of Admixtures are used.

- Hydrophobic or water repellent admixtures which form a water repellent layer along the concrete, but the pores themselves remain open.
- Finely divided solids either inert or chemically active fillers such as talc, clay, silicious powders etc. These densify the concrete and physically limit the water passage through the pores.
- Crystalline Products are proprietary active chemicals in a carrier of cement and sand. These are hydrophilic pore blocking materials that increase the density of calcium silicate or generate crystalline deposits that block pores to resist water penetration.

Some polymer latex products are also used as admixtures that can resist hydrostatic pressure but are unable to bridge cracks and may not produce watertight concrete.

- The Hydrophilic crystalline admixtures are the most successful in providing the best resistance to water under hydrostatic pressure. The active ingredients react with water and cement particles to form calcium silicate crystals that integrally bond with the cement paste. These crystalline deposits block both pores and microcracks in the concrete, to prevent the passage of water. The reaction continues over the life of the concrete, serving to seal not only initial shrinkage cracks but cracks which may occur at a later stage.

3.5 Water Stops

Whenever a construction joint occurs in a basement concrete structure, a water stop should be installed in the joint to prevent the transmission or passage of water through the joint. Construction joints are also called Cold Joints, occur when one section of concrete is laid and cured -fully or partially before the next stage of concrete is laid. This is quite common in.

- Transition between horizontal and vertical sections
- Insufficient formwork
- Change in form design
- Stoppage of concrete work

When the joints are formed, the Construction joints refer to the concrete structure where two different concrete placements meet. This can also be due to formation of Control Joints, which are added to poured-in-place concrete structure to avoid and control cracks due to shrinkage in large placements. These joints are typically the weakest points of concrete.

Water stops are generally recommended to avoid Cold/Concrete/Controlled joints. The capability of Water stops to prevent water ingress at these weak points in the structure is critical to successful waterproofing of basement structures. It is important to incorporate in the Waterproofing system right at the Waterproofing Design stage.

There are different types of Water stops, having different shapes, sizes, and composition. Some of these are.

- PVC (Polyvinyl Chloride)
- Neoprene Rubber
- Thermoplastic Rubber
- Bentonite Clay
- Pre-Formed Hydrophilic Resins

The choice and selection of right Water stop depends upon the building design, ground water level etc. Its important to lay the Water stop in place properly and ensure that it does not move during concrete pouring. It should be noted that the Water stops are not used for Expansion Joints.

Chapter 4

Surface Preparation and Inspection

Effective waterproofing begins with thorough surface inspection, meticulous substrate preparation, and diligent repair of cracks and defects. This chapter delves into the essential aspects of assessing surfaces, preparing substrates, and addressing imperfections to ensure a successful waterproofing application. Divided into three sections, it provides comprehensive guidance on surface inspection, substrate preparation, and repair techniques tailored to different types of surfaces, including concrete walls, floors, and plastered surfaces.

4.1 Surface Assessment

4.2 Proper Substrate Preparation

4.3 Repairing Cracks and Defects

4.1 Surface Inspection

Surface inspection is the first step in identifying potential issues that may compromise the integrity of waterproofing systems. This section outlines what to observe during inspections, essential tools for assessment, inspection methodologies, and key considerations to keep in mind. By conducting thorough inspections, stakeholders can identify existing defects, moisture ingress points, and substrate conditions that may impact waterproofing performance.

4.2 Substrate Preparation

Substrate preparation is crucial for ensuring proper adhesion and performance of waterproofing materials. This section provides detailed instructions on how to prepare surfaces for waterproofing, including surface cleaning, levelling, and priming. It discusses the tools and equipment required for substrate preparation, best practices for surface cleaning, and precautions to take to avoid common issues such as contamination and poor adhesion.

4.3 Repair of Cracks and Defects

Repairing cracks, pinholes, and concrete imperfections is essential for creating a seamless and durable waterproofing system. This section explores various repair techniques using tapes, micro concrete, and polymer-modified mortars. It outlines step-by-step procedures for addressing different types of defects, including corner joints and substrate irregularities. By employing appropriate repair methods, stakeholders can enhance the performance and longevity of waterproofing systems.

Through comprehensive surface inspection, meticulous substrate preparation, and effective repair techniques, stakeholders can mitigate the risks of water ingress and ensure the long-term integrity of waterproofing systems. This chapter equips readers with the knowledge and tools necessary to achieve reliable and durable waterproofing solutions for a variety of surface types and conditions.

Chapter 5

Waterproofing Applications

5.1 Waterproofing Foundations and Basements

The waterproofing of foundations and basements is a critical aspect of construction and building maintenance. In this part of the chapter, we delve into the significance of effectively waterproofing these structural elements to mitigate the risks associated with water infiltration. Foundations serve as the backbone of any structure, bearing the weight of the building and transferring it to the ground. Basements, on the other hand, provide valuable additional space but are particularly vulnerable to water damage due to their below-grade positioning. Therefore, ensuring proper waterproofing measures are in place is essential to safeguarding the integrity and longevity of buildings. Throughout this chapter, we explore various techniques and materials used for waterproofing foundations and basements, considering factors such as water table levels, soil conditions, and structural requirements. By understanding the importance of waterproofing and implementing appropriate strategies, construction professionals can enhance the durability and resilience of buildings against moisture-related issues, ultimately promoting safer and more sustainable built environments.

5.1.1 Waterproofing of Foundations: In construction, various types of foundations and footings are utilized to support and distribute the load of a structure onto the underlying soil or rock. Some common types of footings and their suitable waterproofing systems:

1. Strip Footings: Also known as spread footings, these are continuous footings that support load-bearing walls or columns. They distribute the load over a wider area of soil.

Most suitable waterproofing systems are Liquid-applied membrane or cementitious coating. Strip footings are continuous footings that support load-bearing walls or columns. Liquid membranes and cementitious coatings provide seamless waterproofing protection, accommodating minor movements in the foundation without compromising integrity. They are also easy to apply and adapt to the long, narrow shape of strip footings.

2. Pad Footings: Similar to strip footings, pad footings are isolated footings that support individual columns or points of load. They are often used when the load from a single column is relatively light. Preformed membranes are the most suitable waterproofing systems. Pad footings are isolated footings that support individual columns or points of load. Preformed membranes provide robust waterproofing protection, forming a barrier against water infiltration at specific points of load. They are also durable and resistant to punctures or tears.

3. Raft or Mat Foundations: These are large, continuous concrete slabs that cover the entire footprint of a building. They distribute the load over a large area, making them suitable for buildings constructed on weak or expansive soils. Most suitable waterproofing system is a combination of Crystalline Admixture and HDPE (high-density polyethylene) membrane. Raft or mat foundations cover the entire footprint of a building, requiring a waterproofing method that can withstand expansive soil conditions and provide broad coverage. HDPE membranes offer excellent resistance to water and chemical exposure over large areas.

4. Pile Foundations: Pile foundations consist of long, slender structural elements (piles) driven deep into the ground to transfer the load to a deeper, more stable soil or rock layer. Types of pile foundations include driven piles, bored piles, and screw piles. Cementitious Crystalline grout or epoxy injection are the most suitable for waterproofing of pile foundations. Pile foundations involve deep structural elements driven into the ground to transfer loads. Waterproofing methods such as cementitious crystalline grout or epoxy injection can seal cracks or voids in the pile structure, preventing water ingress and protecting the foundation against corrosion or degradation.

5. Pier Foundations: Pier foundations consist of cylindrical or rectangular vertical members (piers) that support the structure's load. They are often used in areas with deep frost lines or expansive soils. Cementitious coating or spray-applied membrane are among the most suitable waterproofing systems. Pier foundations consist of vertical members supporting the structure's load. Cementitious coatings or spray-applied membranes provide effective waterproofing protection for the exposed surfaces of pier foundations, ensuring durability and longevity.

The choice of foundation type depends on various factors, including the soil type, site conditions, structural requirements, and local building codes and regulations. Each type of foundation and footing has its advantages and limitations, and selecting the appropriate one is crucial for ensuring the stability and longevity of a structure. Different types of footings require different waterproofing methods based on their structural characteristics, load-bearing capacity, and soil conditions. Overall, selecting the appropriate waterproofing method for different types of footings involves considering factors such as the foundation's design, construction materials, soil conditions, and environmental factors to ensure long-term performance and structural stability.

5.1.2 Basement Waterproofing:

Basements typically have various areas that require treatment:

a. Below the Raft:

This refers to the area beneath the concrete raft or slab foundation of the basement. It's crucial to ensure waterproofing in this space to prevent water from seeping up into the basement area. Depending on the specific requirements, various waterproofing methods or a combination can be employed, such as HDPE membrane systems, crystalline admixture, crystalline coatings, Cementitious Crystalline injection grouting, or hydrophilic systems.

b. Retaining Wall from Positive Side:

This involves waterproofing the retaining wall on the side facing the interior of the basement. Positive side waterproofing means addressing the side where water pressure is not present. Common methods include applying waterproof coatings include Crystalline Admixture, Cementitious Crystalline Injection Grouting, Swellable water bars, Liquid Applied Coatings and Preformed Membranes to prevent moisture infiltration. A combination of systems can be used to have a moisture free basements.

c. Retaining Wall from Negative Side:

Negative side waterproofing refers to treating the side of the retaining wall facing the exterior, where water pressure is present. This is often more challenging as it requires addressing the water pressure exerted on the wall. Solutions may include installing drainage systems, using Cementitious Crystalline Admixture, applying crystalline coatings, Cementitious Crystalline Injection grouts to resist water ingress.

Each of these areas requires careful consideration of factors such as water pressure, moisture exposure, and structural conditions to determine the most suitable waterproofing approach. The choice between rubberized, acrylic, or cementitious coatings depends on the specific requirements of the project, including budget, site conditions, and the desired level of waterproofing performance.

5.2 Waterproofing Roofs and Terraces

Effective waterproofing of roofs, terraces, and balconies is crucial for maintaining the structural integrity and longevity of buildings. These horizontal surfaces are constantly exposed to environmental elements such as rain, snow, UV radiation, and temperature fluctuations, making them vulnerable to water ingress and damage. In this chapter, we explore the importance of waterproofing these areas and recommend systems to ensure durable protection against moisture infiltration. Why should we Waterproof the Roofs and Terraces:

1. Structural Integrity: Water penetration can compromise the structural integrity of roofs, terraces, and balconies, leading to deterioration of building materials, including concrete, steel, and wood. Waterproofing helps prevent moisture-induced corrosion, cracking, and weakening of structural elements.
2. Prevents Water Damage: Leakage through roofs, terraces, and balconies can result in water damage to interior spaces, including ceilings, walls, and furnishings. Proper waterproofing mitigates the risk of costly repairs and maintains a healthy indoor environment by preventing mold and mildew growth.
3. Enhances Longevity: Waterproofing extends the lifespan of roofs, terraces, and balconies by protecting them from moisture-related deterioration. By minimizing water intrusion, waterproofing systems contribute to the durability and longevity of building structures.

Recommended Waterproofing Systems:

1. Preformed Membrane Systems:

Preformed membranes such as APP membranes consist of layers of asphalt-modified bitumen reinforced with fiberglass or polyester. These membranes provide excellent waterproofing properties and are suitable for flat or low-slope roofs, terraces, and balconies. They can be torch-applied, self-adhered, or cold-applied, offering flexibility in installation methods.

2. Liquid-Applied Membrane Systems:

Liquid-applied membranes are polymer-based coatings that form a seamless, flexible barrier over the substrate. These systems are highly durable and resistant to UV radiation, making them suitable for exposed roof surfaces, terraces, and balconies. Liquid membranes can be spray-applied or roller-applied, providing easy and efficient coverage.

3. Polyurethane Coating Systems:

Polyurethane coatings offer superior waterproofing and elastomeric properties, allowing for movement of building structures without compromising the integrity of the waterproofing system. These coatings adhere well to various substrates and are resistant to weathering, making them ideal for roof decks, terraces, and balconies exposed to harsh environmental conditions. A hybrid system of Polyurea and Polyurethane coatings can also be used for a high strength system.

4. Cementitious Waterproofing Systems:

Cementitious waterproofing materials, such as cement-based coatings or membranes, provide durable protection against water infiltration. These systems are commonly used for terrace waterproofing and balcony waterproofing due to their ease of application and compatibility with concrete substrates. Cementitious waterproofing products can also be reinforced with additives to enhance their performance and longevity.

In conclusion, waterproofing of roofs, terraces, and balconies is essential for preserving the structural integrity, preventing water damage, and extending the lifespan of buildings. By selecting and installing appropriate waterproofing systems, construction professionals can ensure reliable protection against moisture infiltration, contributing to the overall durability and sustainability of built environments.

5.3 Waterproofing Bathrooms and Wet Areas

When it comes to construction and maintenance of buildings, one of the most important aspects to consider is waterproofing. In particular, wet area waterproofing is a crucial element to protect bathrooms, laundries, balconies and other areas of the building that are constantly exposed to moisture and water.

Wet area waterproofing serves the purpose of preventing water from seeping into undesired areas, which can lead to structural damage to the building. It also ensures that the area is hygienic, as moisture provides an ideal condition for microbial growth.

There are different waterproofing methods for wet areas, but the traditional technique involves installing a waterproof membrane, which acts as a barrier between moisture and the underlying surfaces. This membrane is typically made of materials like asphalt, PVC or bitumen and is applied directly onto the surface before tiling or finishing.

Another popular method is using a liquid waterproofing agent, such as polyurethane or acrylic-based products. These products are brushed, sprayed or rolled onto the surface, creating a seamless, flexible membrane that conforms to the contours of the structure.

It is important to note that waterproofing solutions must be installed by trained professionals with experience in wet area waterproofing, ensuring that it is done correctly and meets required standards. The failure to do so can result in costly repairs and even pose a safety risk to the occupants.

Regular maintenance of the waterproofing system is also crucial, as it can prolong the life of the waterproofing membrane and ensure the longevity of the structure. Proper care includes checking for any cracks, damage or leaks, and making necessary repairs as soon as possible.

Several waterproofing systems are suitable for wet area waterproofing, providing protection against water infiltration in areas such as bathrooms, kitchens, and utility rooms. Some common types of waterproofing systems for wet areas include:

A. Liquid-Applied Membrane Systems: Liquid waterproofing membranes are polymer-based coatings that are applied directly onto the substrate using a brush, roller, or spray equipment. Once cured, they form a seamless, flexible barrier that effectively prevents water penetration. Liquid-applied membranes offer excellent adhesion to various substrates and can accommodate movement and settlement, making them ideal for wet area waterproofing.

B. Cementitious Waterproofing Systems: Cementitious waterproofing materials, such as cement-based coatings or membranes, are mixed with water and applied directly onto the substrate. These systems rely on the hydration of cement particles to form a dense, impermeable layer that resists water penetration. Cementitious waterproofing systems are durable, cost-effective, and suitable for wet areas such as shower stalls and bathroom floors.

C. Bore Packs, Corner Tapes and Grouts: The wet areas have Core Cuts which need to be sealed and packed. A combination of Micro concrete based grouting powders & bore packs, Corner tapes and Pipe collars are useful for sealing the core cuts. The tapes are used to seal the wall and floor joints. Different type of tapes can be used for corner sealing. A combination of tapes and cementitious angle fillets are the best choice.

Each type of waterproofing system has its advantages and limitations, and the selection depends on factors such as the substrate, project requirements, budget, and environmental conditions. By choosing the appropriate waterproofing system for wet areas, construction professionals can ensure effective protection against water damage and maintain the integrity of building structures.

5.4 Waterproofing Podiums

Podium waterproofing is a critical aspect of building construction, particularly in urban landscapes where space constraints lead to multi-functional podiums. These podiums serve various purposes such as recreational areas, gardens, driveways, and car parking spaces. Ensuring their waterproofing is essential to protect the structure from water ingress, which can lead to structural damage, mold growth, and deterioration of finishes. In this chapter, we'll explore the different types of podiums and the waterproofing methods specific to each type.

Types of Podiums:

1. Podiums with Soft scaping:

These podiums typically feature green spaces, gardens, and landscaping elements such as grass, shrubs, and trees. Waterproofing in such podiums must provide protection to the underlying structure while allowing for proper drainage to maintain the health of the vegetation. Waterproof membranes that are root-resistant are often used to prevent damage from plant roots.

2. Podiums with Hardscaping:

Podiums with hardscaping involve the use of non-living elements such as concrete, stone, and paving materials. The waterproofing system in these podiums must ensure protection against water penetration while also providing durability to withstand foot traffic and environmental factors. Liquid applied membranes or sheet membranes reinforced with fabric are commonly used for such applications.

3. Podiums with Driveways & Car Parking:

These podiums endure heavier loads and vehicular traffic, requiring robust waterproofing systems capable of withstanding dynamic loads and chemical exposure from vehicles. Additionally, proper drainage systems are crucial to prevent water accumulation, which can lead to safety hazards and structural deterioration. Waterproofing solutions for these podiums often include multi-layered systems with a combination of membranes, coatings, and drainage mats.

Suggested Waterproofing Methods for Podiums:

1. Membrane Systems:

Membrane systems are commonly used for podium waterproofing due to their flexibility, ease of installation, and effectiveness in providing a continuous barrier against water ingress. These membranes can be made of bitumen, PVC, EPDM, or polymer-modified materials. They are applied in multiple layers with overlapping seams to ensure a watertight seal.

2. Liquid Applied Waterproofing:

Liquid waterproofing membranes offer versatility in application and can conform to irregular surfaces and intricate details. They are often used in conjunction with membrane systems or as standalone waterproofing solutions. Liquid membranes are applied in multiple coats to achieve the desired thickness and coverage, forming a seamless protective barrier.

3. Drainage Systems:

Proper drainage is essential for podiums to prevent water accumulation and hydrostatic pressure build-up. Drainage systems may include channels, pipes, and drainage mats installed beneath the waterproofing layer to redirect water away from the structure. Additionally, slope adjustments and strategically placed drains help facilitate efficient water runoff.

4. Joint Sealants:

Joints and penetrations in podium structures are vulnerable points for water ingress. Sealants specifically designed for waterproofing applications are used to seal joints, gaps, and connections, preventing water infiltration. These sealants accommodate movement and maintain flexibility to ensure long-term effectiveness.

Podium waterproofing requires careful consideration of the podium's function, construction materials, and environmental conditions. By employing appropriate waterproofing methods and materials tailored to the specific requirements of each type of podium, builders can ensure the longevity, durability, and structural integrity of these elevated spaces. Effective waterproofing not only protects the building structure but also enhances the usability and aesthetic appeal of the podium for years to come.

5.5 Waterproofing of Terrace Gardens

In recent years, the trend of terrace and kitchen gardens has been gaining popularity, especially in urban areas where space for traditional gardens is limited. The shift towards terrace gardens is driven by various factors, including the desire for green spaces, the benefits of homegrown produce, and the increased awareness of environmental sustainability. However, developing a terrace garden comes with its own set of challenges, particularly in terms of waterproofing to protect the underlying structure. In this chapter, we'll explore the benefits of terrace and kitchen gardens, the unique waterproofing requirements they pose, and the role of specific waterproofing systems such as Polyurea-Polyurethane hybrid, Pitch Modified Polyurethane, and Pitch Modified Acrylic systems. Additionally, we'll discuss how an anti-root waterproofing system contributes to better waterproofing of terrace gardens.

Benefits of Terrace & Kitchen Gardens:

1. Space Optimization: Terrace gardens utilize unused rooftop space, allowing homeowners to maximize their property's potential and create green areas for relaxation, recreation, and cultivation.

2. Fresh Produce: Terrace and kitchen gardens enable homeowners to grow their own fruits, vegetables, herbs, and

flowers, providing access to fresh, organic produce right at their doorstep.

3. Environmental Benefits: By incorporating greenery into urban landscapes, terrace gardens contribute to improved air quality, reduced heat island effect, and enhanced biodiversity. They also promote water conservation through rainwater harvesting and reduce carbon emissions associated with transporting produce from distant farms.

4. Health and Well-being: Gardening is known to have therapeutic benefits, promoting stress relief, physical activity, and mental well-being. Terrace gardens offer a tranquil retreat within the confines of one's home, fostering a deeper connection with nature.

Waterproofing Requirements for Terrace Gardens:

Developing a terrace garden requires careful consideration of waterproofing to prevent water leakage, seepage, and structural damage. Key waterproofing requirements for terrace gardens include:

1. Protection against Water Penetration: The waterproofing system must provide a seamless barrier to prevent water from infiltrating the underlying structure, including the roof slab and walls.

2. Root Resistance: Since terrace gardens involve the cultivation of plants, trees, and shrubs, the waterproofing system should be resistant to root penetration to avoid damage and maintain structural integrity.

3. Flexibility and Durability: The waterproofing membrane must be flexible enough to accommodate structural movement and temperature fluctuations without cracking or compromising its effectiveness. It should also be durable enough to withstand foot traffic and exposure to UV radiation.

4. Compatibility with Green Roof Components: If incorporating a green roof design, the waterproofing system should be

compatible with drainage layers, growing media, and vegetation to ensure proper functioning and longevity of the garden.

Waterproofing Systems for Terrace Gardens:

1. Polyurea-Polyurethane Hybrid System:

This system combines the fast-curing properties of polyurea with the flexibility and adhesion of polyurethane. It forms a seamless, elastomeric membrane that can withstand harsh weather conditions, UV exposure, and mechanical stress.

2. Pitch Modified Polyurethane System:

Pitch modified polyurethane systems incorporate modified bitumen or pitch to enhance the membrane's elasticity and resistance to root penetration. These systems offer excellent waterproofing properties and long-term durability, making them suitable for terrace gardens.

3. Pitch Modified Acrylic System:

Pitch modified acrylic systems utilize acrylic polymers modified with bitumen or pitch to create a robust waterproofing membrane. They provide a cost-effective solution with good adhesion to various substrates and resistance to weathering.

4. Anti-Root Waterproofing System:

An anti-root waterproofing system incorporates additives or barriers specifically designed to deter root growth and prevent damage to the waterproofing membrane. By inhibiting root penetration, this system ensures the long-term integrity of the terrace garden waterproofing, reducing the risk of leaks and structural issues.

Terrace and kitchen gardens offer numerous benefits, ranging from sustainable food production to environmental conservation and personal well-being. However, ensuring proper waterproofing is essential to protect the underlying structure and maintain the longevity of the garden. By utilizing specialized waterproofing systems such as

Polyurea-Polyurethane hybrids, Pitch Modified Polyurethane, and Pitch Modified Acrylic systems, coupled with anti-root waterproofing measures, homeowners can enjoy the beauty and benefits of terrace gardens

5.6 Waterproofing of External Walls

Waterproofing external walls is crucial for maintaining the structural integrity and longevity of buildings. External walls are constantly exposed to harsh environmental conditions such as rain, humidity, temperature fluctuations, and UV radiation, making them susceptible to water ingress, moisture retention, and deterioration over time. In this chapter, we'll delve into the importance of waterproofing external walls and explore various coatings recommended for effective waterproofing, including Surface Saturating Acrylic Coatings, Anti-Carbonation Coatings in different colours, and Anti-Carbonation Coatings with Thermal Insulation properties.

Importance of Waterproofing External Walls:

1. Protection against Water Damage: Waterproofing external walls prevents water infiltration, which can lead to structural damage, including cracks, efflorescence, spalling, and corrosion of reinforcement steel.

2. Prevention of Mold and Mildew: Moisture accumulation within walls creates an ideal environment for mold and mildew growth, which not only compromises indoor air quality but also poses health risks to occupants.

3. Enhanced Thermal Performance: Waterproofing coatings with thermal insulation properties help regulate indoor temperature by reducing heat transfer through walls, resulting in energy savings and improved comfort.

4. Preservation of Aesthetic Appearance: Waterproofing coatings preserve the aesthetic appeal of external walls by preventing

discoloration, staining, and degradation of finishes caused by water damage and environmental pollutants.

Recommended Waterproofing Coatings:

1. Surface Saturating Acrylic Coatings:

Surface saturating acrylic coatings penetrate deep into the substrate, forming a breathable, protective barrier that repels water while allowing moisture vapor to escape. These coatings provide excellent adhesion, flexibility, and UV resistance, enhancing the durability and longevity of external walls.

2. Anti-Carbonation Coatings in Different Colours:

Anti-carbonation coatings contain additives that inhibit carbon dioxide penetration into the concrete substrate, reducing the risk of carbonation-induced corrosion of reinforcement steel. These coatings are available in a variety of colours, allowing for customization and aesthetic enhancement of external walls while providing long-lasting protection against water ingress and environmental degradation.

3. Anti-Carbonation Coatings with Thermal Insulation Properties:

Anti-carbonation coatings with thermal insulation properties incorporate insulating additives such as ceramic microspheres or reflective pigments to improve the thermal performance of external walls. These coatings help reduce heat transfer, minimize thermal bridging, and enhance energy efficiency, making them ideal for both waterproofing and insulation applications in buildings.

Waterproofing external walls is essential for preserving the structural integrity, aesthetic appearance, and thermal performance of buildings. By employing effective waterproofing coatings such as Surface Saturating Acrylic Coatings, Anti-Carbonation Coatings in different colours, and Anti-Carbonation Coatings with Thermal Insulation properties, property owners can protect their investments and ensure the long-term durability and sustainability of their structures. Additionally, proper maintenance and periodic inspection of waterproofing systems are critical to address any potential issues and prolong the lifespan of external walls in buildings.

5.7 Waterproofing of Swimming Pools

Waterproofing swimming pools is crucial to prevent water leakage, structural damage, and potential safety hazards. The hydrostatic pressure exerted by water can increase the likelihood of waterproofing failure if not properly addressed. In this chapter, we'll discuss the importance of using specific waterproofing systems and products tailored for swimming pools, including the use of cementitious crystalline admixtures, swellable water bars, coatings, and corner tapes.

Hydrostatic Pressure and Waterproofing Challenges:

The constant presence of water in swimming pools subjects the structure to hydrostatic pressure, which can compromise the integrity of waterproofing systems if not adequately addressed. Failure to mitigate hydrostatic pressure can result in water infiltration, cracks, and structural deterioration, leading to costly repairs and safety concerns.

Key Waterproofing Components and Systems:

1. Cementitious Crystalline Admixture:

Incorporating a cementitious crystalline admixture into the concrete mix for the pool shell enhances its waterproofing properties by forming insoluble crystals that block water penetration and seal micro-cracks.

2. Swellable Water Bars:

Swellable water bars are used at joints between the raft and retaining walls to provide a flexible and watertight seal, preventing water ingress and ensuring the integrity of the waterproofing system.

3. Coatings:

Various coatings can be applied to the RCC shell of the swimming pool to provide additional waterproofing protection. Options include two-component cementitious crystalline coatings, two-component cementitious coatings, and polyurea-polyurethane hybrid coatings, each offering different levels of durability and chemical resistance.

4. Deep Penetrating Acrylic Saturating System:

After plastering the RCC shell with high-quality mortar, applying a deep penetrating acrylic saturating system to the plastered surface enhances waterproofing and provides an additional layer of protection against water infiltration.

Additional Waterproofing Measures:

1. Corner Tapes:

Installing corner tapes at wall and floor joints ensures a seamless and watertight transition, preventing water from seeping into vulnerable areas and minimizing the risk of leaks.

2. Sealing Core Cuts:

Core cuts made during construction should be properly sealed using a core cut treatment method outlined elsewhere in the book. This prevents water from penetrating through openings in the structure and maintains the integrity of the waterproofing system.

Waterproofing swimming pools requires careful consideration of the unique challenges posed by hydrostatic pressure and constant exposure to water. By implementing specific waterproofing systems and products such as cementitious crystalline admixtures, swellable water bars, coatings, corner tapes, and core cut treatments, builders can ensure the durability, safety, and longevity of swimming pool structures. Proper installation and maintenance of waterproofing systems are essential to prevent water leakage, protect the integrity of the pool shell, and provide a safe and enjoyable environment for users.

5.8 Waterproofing of Water Storage Tanks

Water storage tanks play a critical role in providing clean and safe water for various purposes, making effective waterproofing essential to maintain water quality and structural integrity. In this chapter, we'll explore the importance of choosing the right waterproofing products for the interior of water storage tanks, including two-component cementitious coatings with high polymer content and certified food-grade epoxy coatings.

The interior surfaces of water storage tanks are constantly exposed to water, making them susceptible to deterioration, corrosion, and contamination if not properly waterproofed. Interior waterproofing ensures that the stored water remains free from impurities, pathogens, and leaks, preserving its quality and safety for consumption.

Recommended Waterproofing Products:

1. Two-Component Cementitious Coating with High Polymer Content:

A two-component cementitious coating with high polymer content offers excellent adhesion, flexibility, and durability, making it an ideal choice for waterproofing the interior surfaces of water storage tanks. The polymer content enhances the coating's resistance to water penetration, chemical exposure, and mechanical stress, ensuring long-lasting protection against leaks and corrosion. Using a Food Grade cementitious coating would be further better.

2. Food-Grade Epoxy Coating:

For water storage tanks intended for potable water use, a certified food-grade epoxy coating provides a safe and hygienic waterproofing solution. These coatings are specifically formulated to meet stringent regulatory standards for contact with food and drinking water, ensuring that the stored water remains free from contaminants and maintains its purity.

Waterproofing the interior surfaces of water storage tanks is essential to maintain water quality, prevent leaks, and ensure the long-term durability of the tanks. By choosing appropriate waterproofing products such as two-component cementitious coatings with high polymer content and certified food-grade epoxy coatings, builders can effectively protect the stored water from contamination and maintain compliance with regulatory standards for potable water storage. Proper surface preparation, application techniques, and curing procedures are crucial for the successful implementation of interior waterproofing systems, providing reliable protection and peace of mind for water storage tank operators and users.

5.9 Waterproofing of Sewage Treatment Plant

Proper waterproofing is crucial for sewage treatment plants in various settings like hotels, group housing, farm houses and institutions. While cementitious or crystalline products can be good primary products for treating the primary structure, its important to use products such as Tar Extended Epoxies or solvent free chemical resistant epoxy on the surface which will come in direct contact with sewage water, due to their effectiveness to their resistance to harsh and aggressive material in sewage water.

5.10 Waterproofing & Treatment of Expansion Joints

In multi-storied buildings and structures with large basements, podiums and terraces, expansion joints play a crucial role in protecting buildings during ground movements such as earth quakes, thermal expansion and structural settlement. To effectively manage varied joint sizes ranging from 40-300mm, a range of flexible joint systems are recommended. For narrow joints, such as around 40-50mm, Polyurethane Sealants, or rubber expansion joints systems provide excellent compression and flexibility. For wider joints, modular expansion joint system featuring a combination of elastomeric inserts such as EPDM or TPO membranes and steel or aluminium profiles offer durability and flexibility to withstand substantial movement. Proper selection and application will ensure structural integrity and longevity in diverse building environments.

Chapter 6

Maintenance and Repair

Maintaining and repairing existing buildings is essential to ensure their longevity, safety, and functionality. In the Indian building context, where structures are subjected to diverse climatic conditions and varying levels of exposure to environmental elements, proper waterproofing maintenance and repair are particularly critical. In this chapter, we'll discuss common waterproofing issues in existing buildings, methods for conducting thorough building inspections for waterproofing, and strategies for repairing and restoring waterproofing systems.

6.1 Common Waterproofing Issues in Existing Buildings

1. Leaking Roofs and Terrace Gardens: Roof and terrace leaks are common issues in existing buildings, often caused by inadequate waterproofing membranes, deteriorated sealants, or improper installation. Water infiltration through these areas can lead to structural damage, mold growth, and interior water damage.

2. Basement Water Seepage: Basement areas are prone to water seepage due to groundwater pressure, poor drainage, or cracks in the foundation walls or floor slab. This can result in dampness, mold, and deterioration of finishes, posing health risks to occupants and compromising the structural integrity of the building.

3. Cracked Walls and Facades: Cracks in exterior walls and facades can allow water ingress, leading to efflorescence, spalling, and staining. These cracks may be caused by settlement, thermal

expansion, or structural movement, requiring proper sealing and waterproofing to prevent further deterioration.

4. Expansion Joint Failure: Expansion joints are susceptible to wear and tear over time, resulting in gaps or failures that allow water penetration. Inadequate joint sealants or improper installation can exacerbate this issue, leading to water infiltration and potential damage to building components.

6.2 Building Inspection for Waterproofing

1. Roof and Terrace Inspection: Check for signs of ponding water, cracks, deteriorated membranes, and damaged sealants on the roof and terrace. Inspect drainage systems, gutters, and downspouts for blockages or defects.

2. Basement Inspection: Examine basement walls and floors for cracks, dampness, efflorescence, and signs of water seepage. Inspect sump pumps, waterproofing membranes, and drainage systems for proper functioning.

3. Exterior Wall Inspection: Look for cracks, gaps, deteriorated sealants, and staining on exterior walls and facades. Pay attention to expansion joints, window sills, and door thresholds for signs of water infiltration.

4. Interior Inspection: Check interior spaces for water stains, mold growth, dampness, and musty odours, indicating potential water ingress. Inspect plumbing fixtures, pipe penetrations, and utility openings for leaks or moisture accumulation.

6.3 Repair and Restoration of Waterproofing

1. Roof and Terrace Repair: Repair damaged membranes, seal cracks, and replace deteriorated sealants to restore waterproofing integrity. Ensure proper slope and drainage to prevent water ponding.

2. Basement Waterproofing: Seal cracks in foundation walls and floor slabs, install interior or exterior waterproofing

membranes, and improve drainage systems to mitigate water seepage issues in basements.

3. Wall and Facade Restoration: Seal cracks, repair damaged mortar joints, and apply waterproof coatings or sealants to exterior walls and facades to prevent water infiltration and enhance durability.

4. Expansion Joint Maintenance: Replace damaged joint sealants, install expansion joint covers or systems, and ensure proper detailing and installation to maintain watertightness and prevent water ingress through expansion joints

Maintenance and repair of existing buildings are essential to address common waterproofing issues and ensure their long-term durability and resilience in the Indian building context. By conducting thorough building inspections for waterproofing, identifying potential issues, and implementing appropriate repair and restoration measures, property owners can protect their investments, enhance occupant comfort, and maintain the integrity of their structures for years to come. Additionally, proactive maintenance programs and periodic inspections are crucial for preventing future waterproofing problems and preserving the

Chapter 7

Key Takeaways and Recommendations

1. Importance of Waterproofing: Waterproofing is essential for protecting buildings from water damage, ensuring structural integrity, and preserving aesthetics. It is a crucial aspect of building construction and maintenance that should not be overlooked.

2. Understanding Waterproofing Techniques: Familiarize yourself with different waterproofing techniques such as membrane waterproofing, liquid applied waterproofing, integral waterproofing, and cementitious waterproofing. Each technique has its advantages and is suited for specific applications.

3. Proper Installation and Maintenance: Proper installation and regular maintenance are essential for the effectiveness of waterproofing systems. Ensure surfaces are adequately prepared before applying waterproofing materials and conduct periodic inspections to identify and address any issues promptly.

4. Addressing Common Issues: Be aware of common waterproofing issues such as roof leaks, basement water seepage, cracked walls, and expansion joint failures. Conduct thorough building inspections to identify potential problem areas and implement appropriate repair and restoration measures.

5. Choosing the Right Products: Select high-quality waterproofing products and materials that are suitable for the specific requirements of your building. Consider factors such as climate,

exposure to environmental elements, and intended use of the structure when choosing waterproofing solutions.

6. Integration of Drainage Systems: Incorporate proper drainage systems into waterproofing designs to manage surface water runoff, prevent water accumulation, and reduce hydrostatic pressure on building components.

7. Importance of Joint Sealants and Expansion Joints: Joint sealants and expansion joints play a critical role in maintaining the integrity of waterproofing systems by accommodating movement, preventing water ingress, and extending the lifespan of building components. Use quality sealants and ensure proper installation to maximize effectiveness.

8. Regular Inspection and Maintenance: Implement a proactive maintenance program that includes regular inspections, maintenance activities, and repairs to address any waterproofing issues promptly and prevent future damage.

9. Investing in Professional Expertise: Consider hiring experienced waterproofing contractors or consultants for complex projects or when in doubt about the best waterproofing solutions for your building. Professional expertise can help ensure the success and longevity of waterproofing systems.

By following these key takeaways and recommendations, property owners, developers, and building professionals can effectively address waterproofing challenges, protect their investments, and maintain the durability and performance of their structures for years to come.

Chapter 8

Embracing Waterproofing Best Practices

8.1 Best Practices for Waterproofing New Buildings

Ensuring effective waterproofing in new buildings is essential to prevent water damage and maintain the structural integrity of the structure. However, overlooking key steps in the waterproofing process can lead to costly repairs and embarrassing situations for building professionals. In this chapter, we'll outline best practices for waterproofing new buildings, covering site inspection, studying drawings and plans, designing waterproofing systems, site pre-inspection, surface preparation, material selection, application, post-application inspection, water testing, coating protection, and client sign-off.

1. Site Inspection:

Begin by conducting a thorough site inspection to assess environmental factors such as water table levels, soil quality, and water quality. Identify potential sources of water ingress and areas prone to moisture accumulation.

2. Study Drawings and Plans:

Review architectural and structural drawings, as well as plumbing and drainage plans, to understand the building layout, potential waterproofing challenges, and critical areas requiring attention.

3. Design Waterproofing Systems:

Based on the drawings and site inspection findings, design waterproofing systems tailored to the specific requirements of the building. Consider factors such as water table fluctuations, soil composition, and anticipated exposure to water.

4. Site Pre-Inspection and Surface Preparation:

Before waterproofing application, conduct a pre-inspection of the site to identify any existing issues or obstacles. Clean and prepare surfaces thoroughly to ensure proper adhesion of waterproofing materials.

5. Material Selection:

Select waterproofing products based on technical data sheets, focusing on material consumption, compatibility with substrates, and performance properties such as flexibility, adhesion, and resistance to environmental factors.

6. Application Process:

Follow manufacturer guidelines for product application, ensuring proper mixing, priming, and application techniques. Use appropriate tools and equipment to achieve uniform coverage and thickness of waterproofing coatings.

7. Post-Application Surface Inspection:

After application, inspect the surfaces for any defects, unevenness, or areas of incomplete coverage. Address any issues promptly to maintain the effectiveness of the waterproofing system.

8. Water Testing:

Conduct water testing to verify the integrity of the waterproofing system and identify any leaks or weaknesses. Apply water at various pressure levels to simulate real-world conditions and ensure thorough waterproofing performance.

9. Coating Protection:

Once the waterproofing system is in place and tested, protect coatings from damage during subsequent construction activities. Implement measures such as protective barriers, signage, and temporary covers to safeguard waterproofing materials.

10. Client Sign-Off:

Finally, obtain sign-off from the client or building owner, documenting the completion of waterproofing works and confirming satisfaction with the results. Provide maintenance guidelines and warranties to ensure ongoing protection of the building.

By following these best practices for waterproofing new buildings, building professionals can mitigate risks, ensure quality outcomes, and avoid embarrassing situations associated with water damage and structural failures. Proper planning, attention to detail, and adherence to industry standards are essential for successful waterproofing projects and the long-term durability of buildings.

8.2 Best Waterproofing Practices for Existing Buildings Requiring Repair

Repairing waterproofing in existing buildings presents unique challenges and requires careful consideration of existing conditions, client expectations, and effective solutions. In this chapter, we'll outline the best practices for waterproofing repairs in existing buildings, covering site inspection, identification of waterproofing failures, client communication, product selection, application, post-application inspection, testing, and client sign-off.

1. Site Inspection:

Begin by conducting a comprehensive site inspection to assess the extent of waterproofing issues. Inspect basements, walls, roofs, and other areas where waterproofing failures are suspected or observed.

Document observations and take measurements as necessary. Ask the client for type of waterproofing done at the time of construction and if any waterproofing repairs were done post construction.

2. Client Communication and Drawing Request:

Communicate with the client to understand their concerns, expectations, and any previous waterproofing attempts. Request architectural drawings or plans to gain insights into the building's construction and potential areas of vulnerability.

3. Identification of Waterproofing Failures:

Investigate the probable causes of waterproofing failures, such as cracks, deteriorated membranes, inadequate drainage, or improper installation. Use diagnostic tools like moisture meters, thermal imaging, and visual inspections to identify problem areas accurately.

4. Safe Dismantling and Disposal:

Plan and execute the safe dismantling and disposal of existing waterproofing materials, ensuring compliance with safety regulations and environmental guidelines. Dispose of debris responsibly and take necessary precautions to minimize disruption to occupants.

5. Product and Technology Selection:

Select appropriate waterproofing products and technologies based on the specific requirements of the repair project. Consider factors such as substrate compatibility, performance properties, durability, and ease of application. Consult with manufacturers or waterproofing experts for recommendations.

6. Application Process:

Follow manufacturer guidelines and industry best practices for product application, ensuring proper surface preparation, priming, and application techniques. Use suitable tools and equipment to achieve uniform coverage and optimal adhesion.

7. Post-Application Inspection and Testing:

Conduct thorough post-application inspections to verify the effectiveness of the repair work. Inspect surfaces for defects, irregularities, or areas of incomplete coverage. Perform water testing to simulate real-world conditions and confirm the integrity of the waterproofing system.

8. Client Sign-Off:

Obtain client sign-off upon completion of the repair work, documenting the successful resolution of waterproofing issues and confirming client satisfaction. Provide warranties, maintenance guidelines, and recommendations for ongoing care to ensure long-term protection.

By following these best practices for waterproofing repairs in existing buildings, contractors and building professionals can address waterproofing issues effectively, restore structural integrity, and

Glossary of Waterproofing Terms

1. Adhesion: The bonding ability of a waterproofing material to the substrate surface.
2. Bitumen: A viscous, black, or dark brown material used in waterproofing applications, derived from petroleum or natural asphalt.
3. Cementitious Waterproofing: Waterproofing method that involves the application of cement-based coatings or slurries to create a protective barrier against water penetration.
4. Efflorescence: White, powdery deposits that form on the surface of masonry materials due to the migration of soluble salts.
5. Expansion Joint: A joint designed to accommodate thermal expansion and contraction of building materials, preventing cracks and damage caused by movement.
6. Hydrostatic Pressure: Pressure exerted by standing or flowing water against a surface, which can lead to water infiltration and structural damage.
7. Integral Waterproofing: Method of incorporating waterproofing agents or admixtures directly into concrete or mortar mixes to enhance water resistance.
8. Membrane Waterproofing: Waterproofing technique involving the application of flexible sheets or membranes onto surfaces to create a watertight barrier.

9. Polyurea: A type of fast-curing, elastomeric waterproofing material known for its high flexibility, durability, and resistance to chemicals and abrasion.
10. Polyurethane: A versatile polymer used in waterproofing applications, known for its flexibility, adhesion, and resistance to UV radiation and weathering.
11. Root Barrier: A waterproofing system designed to prevent root penetration into structures, protecting against damage and maintaining structural integrity.
12. Sealant: A flexible material used to seal joints, gaps, and connections in waterproofing systems, preventing water infiltration and ensuring watertightness.
13. Sump Pump: A device used to remove water from basements or low-lying areas, preventing water accumulation and reducing the risk of flooding.
14. Thermal Bridging: The transfer of heat through building materials, leading to energy loss and potential condensation issues in waterproofing systems.
15. Waterproofing Membrane: A thin layer of material applied to surfaces to prevent water penetration and protect against moisture damage.
16. Weep Hole: A small opening or cavity designed to allow water to drain or escape from behind a waterproofing system, preventing water build-up and pressure.

Standard Specifications

1. Specifications for Terrace Waterproofing

Primary Coating Material: Deep Penetrating & Surface Saturating Acrylic Primer (A) & Elastomeric Single Component, Below Screed Acrylic Waterproofing Membrane (B)

Ancillary Materials: Ultra Flexible 200% Acrylic Cementitious Waterproofing Coating, SBR Modified Polymer, Low viscosity Epoxy **Grout,** Structural Grade Repair Mortar, Cement & Sand

Tools & Tackles: Angle Grinder, Cutting Machine, Grinding Cup, Hammer, Chisel, Grouting Pump, Nozzle, Paddle mixer and Application Roller

METHOD STATEMENT:

SURFACE PREPARATION:

Mechanically Grind the entire surface of roof to remove any contaminants such as dirt and laitance by using a grinding machine with an attachment of brush and If any cracks observed after cleaning mark those cracks and cut open the cracks by making grove so that the surface must be strong & free from dirt, dust & remove loose particles. Clean thoroughly by wire brush & Jet water wash.

CRACK TREATMENT ON TERRACE AREA:

Fine Cracks if any are made open by routing-out to a minimum groove and bond coat of acrylic filler is applied and is sealed with **Crack Sealer.** Deep cracks on the roof are treated with **Low Viscous Epoxy free flow**

grout by manually pouring or injection grouting, the product in the groove.

CONSTRUCTION JOINTS ON TERRACE AREA:

- All construction joints are made open by cutting machine to a minimum ¾" x ¾ "groove.
- **SBR/Acrylic Modified Polymer (C)** to be applied in the above prepared joints before with PM repair mortar.
- When bond coat is tacky it is sealed with site prepared polymer modified mortar OR with **Structural Grade Repair Mortar** workable PMM.

COVING AT THE JUNCTION JOINTS:

- The interface joint of the floor and wall will cut will be open where ever required and clean the surface, giving one bond coat and packed with the site prepared modified mortar a specialized non – shrink cementitious compound with a ratio of 1:4 CM.
- The coving will be done as curve or taper as per the site conditions with 2" x 2" approx... with **SBR/Acrylic Modified Polymer.**

CONSTRUCTION / STARTER JOINTS AT SLAB / PARAPET WALL JUNCTION:

Sealing the starter /construction joints at roof slab / parapet wall joints with **3-PLY TPE** 300mm elastomeric tape sandwiched with **premix Cementitious adhesive.**

Waterproofing Application:

- Dilute **Surface Saturating Acrylic Primer (A)** with water in proportion of 1:3 and stir well until a uniform consistency is achieved. Apply a single coat of diluted **Primer (A)** on the cleaned substrate & allow it to dry for 6 to 7 hours before commencing subsequent waterproofing coating system.

- **Apply Elastomeric Below Screed Acrylic Waterproofing Membrane (B),** a single component liquid applied waterproofing membrane of acrylic co-polymer with resistance to water permeability. The seamless elastomeric membrane performs and adopts to any construction requirement in varied climatic conditions.

- The product has Tensile Strength > 1.5 N/mm2 & Elongation at break of >250% as per ASTM D 412. It should be applied in 2 coats by using a stiff brush, roller or spray to obtain the desired thickness on horizontal and Vertical surfaces with a material consumption of approx.1.2-1.3 Kg/m^2 for two coats with an interval of 6 -7 hrs based on the site condition.

Package and Mixing Ratio:

Primer (A) Consisting of 20 KG Package:

Liquid component 20 kg

Water component 60 kg

Elastomeric Below Screed Acrylic Waterproofing Membrane (B) Consisting of Single Component 20 Kg Package

Recommended Waterproofing System
1.   **Single Coat Surface Saturating Acrylic Primer (A)**
2.   **Two Coats Elastomeric Below Screed Acrylic Waterproofing Membrane (B)**

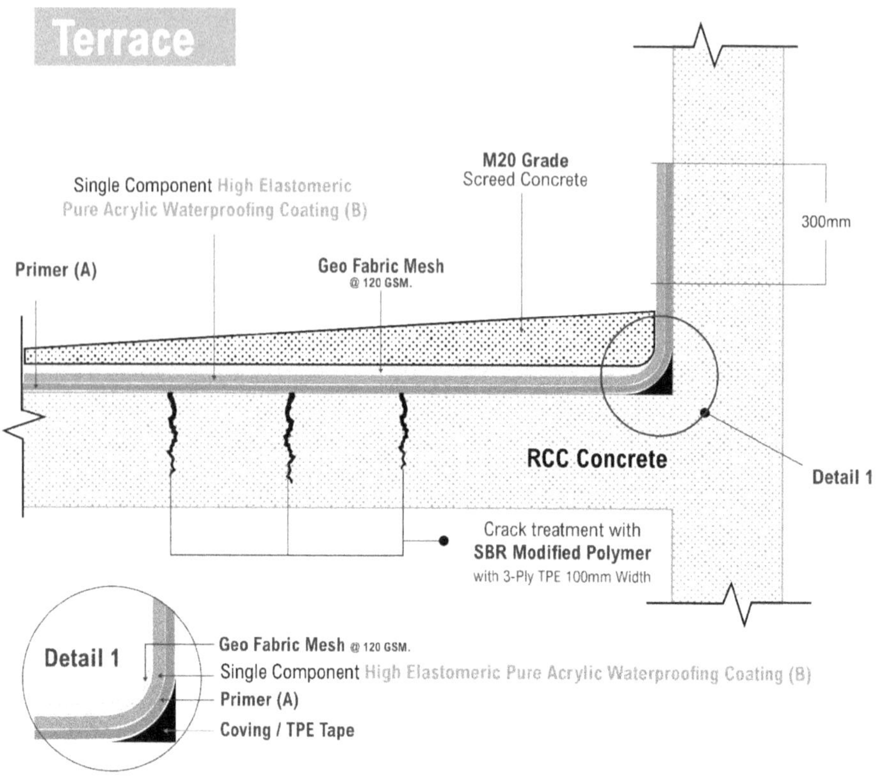

Protection after Waterproofing Application:

Providing protective plaster in **Terrace area** with C.M 1:4 of thickness 15 to 20mm on walls and average thickness of 75mm screed concrete with **SBR/ Acrylic Modified Polymer** over the waterproofing coated surface.

Curing time has to be carried out for minimum 3-4 days

BOQ

S.NO	Descriptions	Unit	Qty.	Rate	Amount
1.	Priming: Supply & Apply water based surface saturation primer composed of Acrylic emulsion polymer, selected special additives in water as a medium. The product has a Density of 1.01 g.cc which helps it saturate the surface it is applied upon. Apply after blending in clean water in 1:3 ratio to be sprayed using HVLP Spray Equipment. To be applied at diluted mix of 12sqm/kg. **As per Code DS-02- SatP-DryShield** Waterproofing				
2.	Supply & Apply a high performance, liquid applied waterproofing membrane of acrylic co-polymer with UV resistance. It is an advanced acrylic technology product manufactured by combination of pigments, reinforced fillers making the polymer flexible that allows the membrane to accommodate movement in the substrate, developed specifically for the **BELOW SCREED** application. Its a seamless membrane that performs and adopts to any construction requirement in varied climatic conditions To be applied in **2 COATS** @ 1.2 kg./sqm to achieve DFT of 1mm2 Elongation at break >250%, Maximum-tensile Strength> 4.5 N/mm2, Shore Hardness 55-60, applied in 2 **coats@1.2kg./sq.mtr.** The coating shall be continued to the entire horizontal area and should be terminated up to parapet top as per **DS-06-WTA-LAP DryShield Waterproofing.**	M^2			
			Total Qty		

2. Waterproofing of Toilet, Balcony & Wet Areas

Primary Coating Material: Surface Saturating Acrylic Repellent Primer (A) & 2 Component Acrylic Cementitious Waterproofing Coating (B)

Ancillary Materials: SBR/Acrylic Modified Polymer, Structural Grade Repair Mortar, Cement & Sand

Tools & Tackles: Angle Grinder, Cutting Machine, Grinding Cup, Hammer, Chisel, Grouting Pump, Nozzle, Paddle mixer and Application Roller

Method Statement:

SURFACE PREPARATION:

Mechanically Grind the entire surface of roof to remove any contaminants such as dirt and laitance by using a grinding machine with an attachment of brush and If any cracks observed after cleaning mark those cracks and cut open the cracks by making grove so that the surface must be strong & free from dirt, dust & remove loose particles. Clean thoroughly by wire brush & Jet water wash.

CRACK TREATMENT:

Fine Cracks if any are made open by routing-out to a minimum groove and bond coat of acrylic filler is applied and is sealed with **Crack Sealer.** Deep cracks on the roof are treated with **Low Viscous Epoxy free flow**

grout by manually pouring or injection grouting, the product in the groove.

CONSTRUCTION JOINTS:

- All construction joints are made open by cutting machine to a minimum ¾" x ¾ "groove.
- **SBR/Acrylic Modified Polymer (C)** to be applied in the above prepared joints before with PM repair mortar.
- When bond coat is tacky it is sealed with site prepared polymer modified mortar OR with **Structural Grade Repair Mortar** workable PMM.

COVING AT THE JUNCTION JOINTS:

- The interface joint of the floor and wall will cut will be open wherever required and clean the surface, giving one bond coat and packed with the site prepared modified mortar a specialized non – shrink cementitious compound with a ratio of 1:4 CM.
- The coving will be done as curve or taper as per the site conditions with 2" x 2" approx... with **SBR/Acrylic Modified Polymer.**

Waterproofing Application

- Dilute **Surface Saturating Acrylic Primer (A)** with water in proportion of 1:3 and stir well until a uniform consistency is achieved. Apply a single coat of diluted **Primer (A)** on the cleaned substrate & allow it to dry for 6 to 7 hours before commencing subsequent waterproofing coating system.
- Apply **Single Component Elastomeric, Acrylic based Liquid Applied Waterproofing Membrane.** The Single component system has to be mixed water in the same proportion as recommended by manufacturer and should be applied in 2 coats by using a stiff brush, roller or spray to obtain the desired thickness on horizontal and Vertical surfaces with a material

consumption of approx..1.1-1.2 Kg/m^2 for two coats with an interval of 6 -7 hrs based on the site condition.

The recommended system has Max. Tensile Strength of >4.5 N/mm2 and Elongation at Break of > 150%. Ideal for small areas up to 1000 sq.ft.

Package and Mixing Ratio:

Primer (A) Consisting of 20 KG Package:

Liquid component 20 kg

Water component 60 kg

Single component Acrylic Coating (B) Consisting of 27 Kg Package:

Recommended Waterproofing System
1. Primer: Single coat Surface Saturating Acrylic Primer (A)
2. Application: Two Coats Application with Single Component Acrylic Liquid Applied Membrane (B)

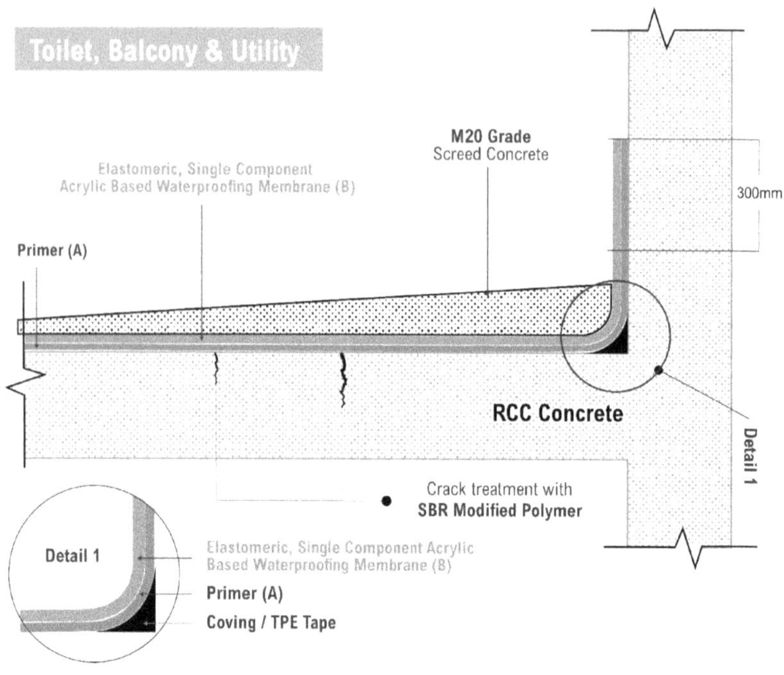

Protection After Waterproofing Application

Providing protective plaster in **Sunken area** with C.M 1:4 of thickness 15 to 20mm on walls with Polymer Modified Mortar over the waterproofing coated surface up to sunken height and coating up to 300 mm height from FFL with sand sprinkling for better tile gripping.

Curing and Water Pond testing for minimum 3 days

BOQ

Sl No	Description	Unit	Qty	Rate	Amount
	Primer: Supply & Apply water based surface saturation primer composed of Acrylic emulsion polymer, selected special additives in water as a medium. The product has a Density of 1.01 g.cc which helps it saturate the surface it is applied upon. Apply after blending in clean water in 1:3 ratio to be sprayed using HVLP Spray Equipment. To be applied at diluted mix of 12sqm/kg. **As per Code DS-02- SatP-DryShield** Waterproofing				
	Supply & Apply **a single component elastomeric liquid applied waterproofing membrane of hybrid acrylic co-polymer. It is an advanced acrylic technology product manufactured by combination of pigments, reinforced fillers making the polymer flexible that allows the membrane to accommodate movement in the substrate. Its a seamless membrane that performs and adopts to any construction requirement in varied climatic conditions. Colour white, Non UV stabilized, Max. Tensile strength >4.5N/mm^2 as per ASTM D 412, Elongation at break >150%, having solid content >65% (ASTM 2369) to be applied@1.1 kg/sq.mtr** on pre saturated surface as per DS-05-WTA-LAP **DryShield Waterproofing.**				
				Total Qty	

3. Waterproofing of UG Water Storage Tank

Nature of Work: Primary Waterproofing on Concrete Surface.

Secondary Protection: Solvent Free Epoxy Coating on Protective Plaster.

Primary Coating Material: Primer (A) & Acrylic Cementitious food grade Waterproofing Coating (B)

Ancillary Materials: Polymer Modified Mortar (C), Cementitious Crystalline Injection Grout, Structural Grade Repair Mortar, Cement & Sand

Secondary Coating Material: Solvent free Food Grade Epoxy Coating Material

Tools & Tackles: Angle Grinder, Cutting Machine, Grinding Cup, Hammer, Chisel, Grouting Pump, Nozzle, Paddle mixer and Application Roller

METHOD STATEMENT:

The area where waterproofing system needs to be installed shall be made free from all other activities that could disrupt the installation of the system.

SURFACE PREPARATION:

Mechanically Grind the entire surface of roof to remove any contaminants such as dirt and laitance by using a grinding machine with an attachment of brush and If any cracks observed after cleaning mark those cracks and cut open the cracks by making grove so that the surface must be strong & free from dirt, dust & remove loose particles. Clean thoroughly by wire brush & Jet water wash.

CRACK TREATMENT:

Fine Cracks if any are made open by routing-out to a minimum groove and bond coat of acrylic filler is applied and is sealed with **Crack Sealer.** Deep cracks on the roof are treated with **Low Viscous Epoxy free flow grout** by manually pouring or injection grouting, the product in the groove.

CONSTRUCTION JOINTS:

- All construction joints are made open by cutting machine to a minimum ¾" x ¾" groove.
- Fill PVC nozzle at 0.5 to 1.0 c/c and inject **with Pre Mixed Cementitious Crystalline Injection Grout.**
- **Acrylic Modified Polymer** to be applied in the above prepared joints.
- When bond coat is tacky it is sealed with site prepared **polymer modified mortar** OR with **Structural Grade Repair Mortar** workable PMM.
- All construction joints are treated with **3-PLY TPE 75 /100mm** elastomeric tape sandwiched with **Acrylic Cementitious food grade Waterproofing Coating (B)**

COVING AT THE JUNCTION JOINTS:

- The interface joint of the floor and wall will cut will be open where ever required and clean the surface, giving one bond coat and packed with the site prepared modified mortar a specialized non – shrink cementitious compound with a ratio of 1:4 CM.

- The coving will be done as curve or taper as per the site conditions with 2" x 2" approx... with **SBR/Acrylic Modified Polymer.**

STARTER JOINTS:

Install a coving on all right-angle bends i.e., Horizontal and Vertical joints. Coving shall be done with Polymer Modified Mortar OR Structural Grade Repair Mortar, as a coving of size 50 mm X 50 mm at the corners of floor and the vertical surface wall area. Sealing joints with **3-PLY TPE TAPE** sandwiched with of **Polymer Modified Mortar OR Cementitious Adhesive**

Grouting:

After Surface Preparation completion ponding test shall be done for 24hrs, if found any leak point shall be arrest by pressure grouting method with **Premixed Cementitious Crystalline Injection Grouting System** from positive side.

Recommended Waterproofing System
1. **Primer: Single coat Primer (A)**
2. **Application: Two Coats Application with Acrylic Cementitious food grade Waterproofing Coating (B)**
3. **Protective Plaster**
4. **Two Coats of Solvent Free Food Grade Epoxy Coating (C)**

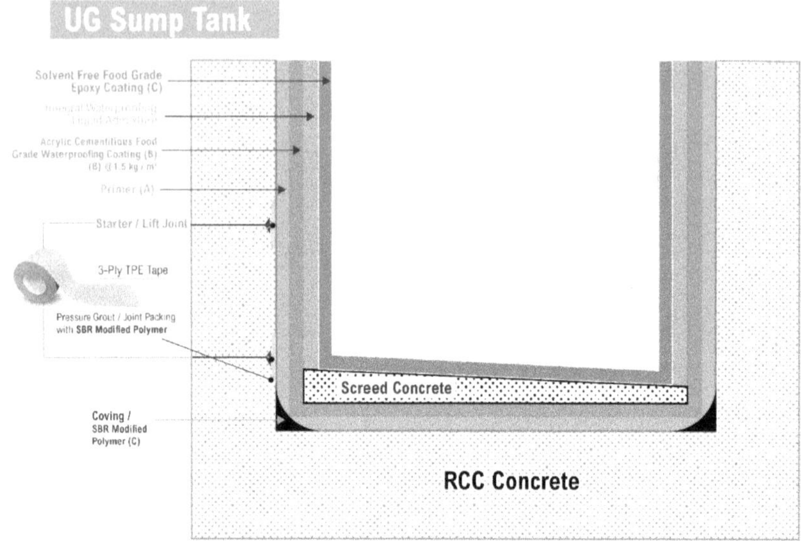

Waterproofing Application:

Primary Waterproofing:

- Dilute **Acrylic Primer (A)** with water in proportion of 1:3 and stir well until a uniform consistency is achieved. Apply a single coat of diluted **Primer (A)** on the cleaned substrate & allow it to dry for 6 to 7 hours before commencing subsequent waterproofing coating system.

- Application of **Acrylic Cementitious Waterproofing Coating (B)** two component elastomeric acrylic cementitious Fibre reinforced waterproof coating Over the Primed surface. The Two component system Part A: two-component Cementitious Waterproofing System Powder and Part B: **Acrylic Cementitious Waterproofing Coating (B)** liquid has to be mixed in the same proportion as recommended by manufacturer. Application should be by using a stiff brush, roller or spray to obtain the desired thickness on horizontal and Vertical surfaces with a material consumption of approx.1.5-1.6 Kg/m² for two coats with an interval of 6 -7 hrs based on the site condition.

Package and Mixing Ratio:

Primer (A) Consisting of 20 KG Package:

Liquid component 20 kg
Water component 60 kg

Acrylic Cementitious Waterproofing Coating (B) Consisting of 30 Kg Package:

Part A: Powder component 20 kg
Part B: Liquid Acrylic Polymer component 10 kg

Note: No extra water to be added.

- **Protection Plaster after Waterproofing Application:** Providing protective plaster in wall with C.M 1:4 of thickness 15 to 20mm on walls **SBR/Acrylic Polymer Modified Mortar** over the waterproofing coated surface up to coated surface height. Floor has to be done screed concrete with M20 grade with proper slope.

Curing time has to be carried out for minimum 3-4 days

Secondary Waterproofing: Food Grade Epoxy

Once the system is 100% dry a surface treatment will be carried out by using a solvent free Epoxy Resin Coating for potable water retaining structures. with two coats using **Solvent Free Epoxy Coating.** The two component should be Food Grade and used for potable water retaining structures and surfaces subject to contact with food stuffs. The cured film is resistant to corrosion, chemical attack and abrasion and is suitable for application in reservoirs, tanks, silos, water treatment works etc. The cured film is non-toxic and meets the requirements of IS:9833-1981.

Package and Mixing Ratio:

Solvent Free Epoxy Coating - Part A: 5kg, Part B: 1 kg

BOQ

S.NO	Descriptions	Unit	Qty.	Rate	Amount
1.	Supply & Apply water based surface saturation primer composed of Acrylic emulsion polymer, selected special additives in water as a medium. The product has a Density of 1.01 g.cc which helps it saturate the surface it is applied upon. Apply after blending in clean water in 1:3 ratio to be sprayed using HVLP Spray Equipment. To be applied at diluted mix of 12sqm/kg. **As per Code DS-02- SatP-DryShield** Waterproofing	M²			
2	Supply & Apply a two component elastomeric coating waterproofing system based on cementitious binders reinforced with micro fibres, special additives, fine-grained selected aggregates, high grade acrylic polymers to provide superior water protection. The Powder: Polymer ratio 1:1, Elongation >200%, Tensile Strength (N/mm2) 0.82, Water Absorption ASTM D 570 1%, Static Crack bridging ability (mm) >2, Shore A Hardness 70.. **The coating should be applied on the pre saturated surface @ 1.4kg./sqm to achieve a DFT of 1mm. As per Dryshield Code DS/2KCemt/ DS 03 DryShield Waterproofing**				
3.	Supply & Apply, a two pack, solvent free, epoxy resin material on the prepared surface to provide Food grade waterproofing coating for potable water. The product has a density of 1.4 g/cc having excellent bond strength, DFT of 200 microns and gets fully cured in 7 days @30 deg. C. The coating has a coverage of 9-10m2/6 kg pack at 400 micron thickness in 2 coats. as per DS-32-FGEP DryShield Waterproofing	M²			

4. Waterproofing Specification for Podium Driveway Area

Nature of Work: Constructive Grade PU Waterproofing Membrane (B)

Primary Coating Material: Constructive Grade PU Waterproofing Membrane (B)

Ancillary Materials: SBR/Acrylic Modified Polymer (C), Cementitious Crystalline Grout, Structural Grade Repair Mortar, Cement & Sand

Tools & Tackles: Angle Grinder, Cutting Machine, Grinding Cup, Hammer, Chisel, Grouting Pump, Nozzle, Paddle mixer and Application Roller

Method Statement:

SURFACE PREPARATION:

Mechanically grind the entire surface of roof to remove any black fungus material and also the loosely laid cement particles by using a grinding machine with an attachment of brush and if any cracks observed after cleaning mark those cracks and cut open the cracks by making grove so that the surface must be strong & free from dirt, dust & remove loose particles. Clean thoroughly by wire brush & Jet water wash.

CRACK TREATMENT:

Fine Cracks if any are made open by routing-out to a minimum groove and bond coat of acrylic filler is applied and is sealed with **Crack Sealer.**

Deep cracks on the slab are treated with **Low Viscous Epoxy free flow grout** by manually pouring or injection grouted in the groove.

CONSTRUCTION JOINTS:

- All construction joints are made open by cutting machine to a minimum ¾" x ¾ " groove.
- **SBR/Acrylic Modified Polymer** to be applied in the above prepared joints before with PM repair mortar.
- When bond coat is tacky it is sealed with site prepared polymer modified mortar OR with **Structural Grade Repair Mortar** workable PMM.
- All construction joints are treated with **3-PLY TPE 75 /100mm** elastomeric tape sandwiched with **Cementitious Adhesive**

COVING AT THE JUNCTION JOINTS:

- The interface joint of the floor and wall will cut will be open wherever required and clean the surface, giving one bond coat and packed with the site prepared modified mortar a specialized non – shrink cementitious compound with a ratio of 1:4 CM.
- The coving will be done as curve or taper as per the site conditions with 2" x 2" approx... With polymer modified mortar with **SBR/Acrylic Modified Polymer.**

CONSTRUCTION / STARTER JOINTS @ SLAB / PARAPET WALL JUNCTION:

Sealing the starter /construction joints at roof slab / parapet wall joints with **3-PLY TPE** 300mm elastomeric tape sandwiched with **Cementitious Adhesive**

BORE PACKING:

- Chipping around the Rainwater down take Pipes.
- All pipe prepared joints to be grouted with non-shrink cementitious flowable grout mix.

GROUTING:

- If required depending on the site condition after pond testing grouting will be done by using hand operated grout machine.
- Our cementitious grouting consists of **factory packed Cementitious Crystalline grout.**

Waterproofing Application

1. Before application of Polyurethane Liquid Membrane coating surface should be 100% dry condition or moisture free. (Moisture to be checked with moisture meter and should not be more than 5%).
2. First Priming has to be carried out by using single component Acrylic surface saturating flexible primer, coverage typically is 350Gms per m^2/ coat on concrete (Typically 1kg for $35m^2$).
3. Allow Primer aliphatic UV resistance primer to dry for 5 to 8 Hours.
4. Application of First coat of **Constructive Grade PU Membrane (B)** @ 0.65 to 0.75kg/m^2/coat High Performance Polyurethane Liquid membrane will be applied by roller on the primed surface
5. Applied First coat of **Constructive Grade PU Membrane (B)** @ 0.65 to 0.75kg/m^2/coat should be allowed it to air curing for about 10 – 12 hrs interval for the second coat.
6. After every coat checking / clearance given by client engineer before proceeding to next coat.
7. Consumption of the material with Two coats of **Constructive Grade PU Membrane (B)** will be around 1.6 Kg/M^2 to achieve 1.2mm thickness.
8. The system has to be left to atmospheric curing for 5-7 days.
9. Laying of Protective Non oven Geo Fabric of 150 GSM over the membrane

Typical coverages as per desired thickness

Rate of uses			
	Kg/m²		
	Standard	At Least	At Most
1 Coat	0.68	0.65	0.75
2 Coats	0.68	0.60	0.75
Total 2 Coats	1.35	1.20	1.50

Thickness (micron)		
	At Least	At Most
1 Coat	500	600
2 Coats	500	600
Total 2 Coats	1000	1200

Protection after Waterproofing Application:

Providing and laying a M20 Grade Screed concrete as per desired thickness for floor with **Crystalline Concrete Admixture** over the waterproofing coated surface.

Finishing the surface with trowel and making a thread impression of 600 mm x 600mm

Dosage: 1% of **Crystalline Admixture** per 50kg of OPC cement.

Curing for 7days

Special Note:

1. Air curing of the PU will be depending on the humidity conditions at the site during the application time will be
2. considered not on TDS doc...
3. Thickness of the film after a primer followed by 2 coats may vary depends on the concrete surface...
4. Application of PU coating to be avoided during peak monsoon season because of minimum curing period is recommended by a PU chemistry.

Recommended Waterproofing System
1. **Primer: Single coat PU Primer (A) with Constructive Grade PU Waterproofing Membrane (B)**
2. **Application: Two Coats Application with Constructive Grade PU Waterproofing Membrane (B)**

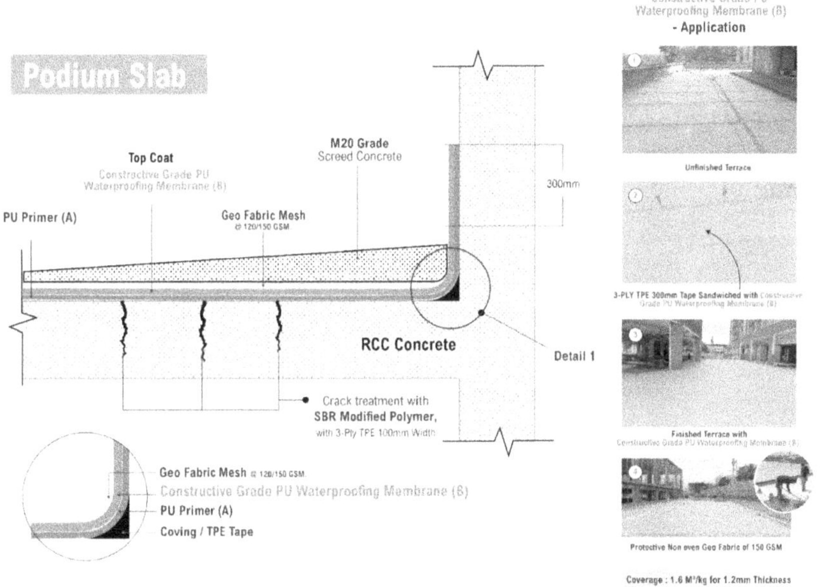

BOQ

S.NO	Descriptions	Unit	Qty	Rate	Amount
1.	Supply & Apply water based surface saturation primer composed of Acrylic emulsion polymer, selected special additives in water as a medium. The product has a Density of 1.01 g.cc which helps it saturate the surface it is applied upon. Apply after blending in clean water in 1:3 ratio to be sprayed using HVLP Spray Equipment. To be applied at diluted mix of 12sqm/kg. **As per Code DS-02- SatP-DryShield** Waterproofing	Sft			
2	Supplying & Applying Providing Waterproofing treatment with **Constructive Grade PU waterproofing Membrane @ 1.6 Kg / m² with** a best quality of single component followed with one coat of primer and all construction joints are cut open in V groove and packed with **Structural Grade fibre based Repair Mortar** and during tacky condition applied with **3- Ply TPE Tape.** Followed by a bond coat as per the manufacturer specifications and Including surface preparation, required all chemicals, tools, labour, supervision, etc. surface preparation tools, labour, joint treatments, supervision etc. as per DS-34-Pod-DryShield Waterproofing	Sft			
4.	Laying a Non oven Geotextile before screed concrete by loosely laying.				
				Total	

5. Waterproofing Specification for Podium – Landscape

MATHOD STATEMENT:

Nature of Work: High Performance Bituminous Modified PU Waterproofing Membrane (B)`

Primary Coating Material: Surface Saturating Acrylic Primer (A) & High-Performance Bituminous Modified PU Waterproofing Membrane (B)

Ancillary Materials: SBR/ Acrylic Modified Polymer, Crystalline Injection Grout, Structural Grade Repair Mortar, Cement & Sand

Tools & Tackles: Angle Grinder, Cutting Machine, Grinding Cup, Hammer, Chisel, Grouting Pump, Nozzle, Paddle mixer and Application Roller

SURFACE PREPARATION:

Grind the entire surface of roof to remove any black fungus material and also the loosely laid cement particles by using a grinding machine with an attachment of brush and If any cracks observed after cleaning mark those cracks and cut open the cracks by making grove so that the surface must be strong & free from dirt, dust & remove loose particles. Clean thoroughly by wire brush & Jet water wash.

CRACK TREATMENT ON PODIUM SURFACE:

Fine Cracks if any are made open by routing-out to a minimum groove and bond coat of acrylic filler is applied and is sealed with **Acrylic**

Sealer. Deep cracks on the roof are treated with **Low Viscous Epoxy free flow grout** by manually pouring or injecting in the groove.

CONSTRUCTION JOINTS ON PODIUM SLAB:

- All construction joints are made open by cutting machine to a minimum ¾" x ¾ "groove.
- **SBR/Acrylic Modified Polymer** to be applied in the above prepared joints before with PM repair mortar.
- When bond coat is tacky it is sealed with site prepared polymer modified mortar OR with **Structural Grade Repair Mortar** workable PMM.
- All construction joints are treated with **3-PLY TPE** 75 /100mm elastomeric tape sandwiched with **High Performance Bituminous Modified PU Waterproofing Membrane (B)**

COVING AT THE JUNCTION JOINTS:

- The interface joint of the floor and wall will cut will be open where ever required and clean the surface, giving one bond coat and packed with the site prepared modified mortar a specialized non – shrink cementitious compound with a ratio of 1:4 CM.
- The coving will be done as curve or taper as per the site conditions with 2" x 2" approx... With polymer modified mortar with **SBR Modified Polymer**

CONSTRUCTION / STARTER JOINTS AT SLAB / PARAPET WALL JUNCTION:

Sealing the starter /construction joints at roof slab / parapet wall joints with **3-PLY TPE** 300mm elastomeric tape sandwiched with **High Performance Bituminous Modified PU Waterproofing Membrane (B)**

BORE PACKING:

- Chipping around the Rainwater down take Pipes.

- All pipe prepared joints to be grouted with non-shrink cementitious flowable grout mix.

Pond Testing has to be carried out for a period of 72 hrs

GROUTING:

- If required depending on the site condition after pond testing grouting will done by using hand operated grout machine.
- Our cementitious crystalline grouting consist of **factory packed premix Crystalline Injection Grout`**

Waterproofing Application

1. Before application of Polyurethane Liquid Membrane coating surface should be 100% dry condition or moisture free. (Moisture to be checked with moisture meter and should not be more than 5%).
2. Apply one coat of **Surface Saturating Acrylic Primer (A)** diluting (1:3) on mother concrete.
3. Apply base coat of with **High-Performance Bituminous modified PU Waterproofing Membrane (B)** with 1:1 dilution.
4. Application of First top coat of with **High-Performance Bituminous Modified PU Waterproofing Membrane (B)** @ 0.65 to 0.75kg/m²/coat High Performance Polyurethane Liquid membrane will be applied by roller on the primed surface
5. Applied First top coat of with **High-Performance Bituminous Modified PU Waterproofing Membrane** (B) @ 0.65 to 0.75kg/m²/coat should be allowed it to air curing for about 10 – 12 hrs interval for the second coat.
6. After every coat checking / clearance given by client engineer before proceeding to next coat.
7. Consumption of the material with Two coats of with **High Performance Bituminous Modified PU Waterproofing Membrane (B)** will be around 1.6 Kg/M² to achieve 1.2mm thickness.

8. The system has to be left to atmospheric curing for 5-7 days.

9. Laying of Protective Non oven Geo Fabric of 150 GSM over the membrane

<table>
<tr><td align="center">Recommended Waterproofing System</td></tr>
<tr><td>

1. **Primer: Single coat Surface Saturating Acrylic Primer (A)**
2. **Application: Two Coats Application with High-Performance Bituminous Modified PU Waterproofing Membrane (B)**
3. **Protect with Geo-Textile followed by Protection Screed**

</td></tr>
</table>

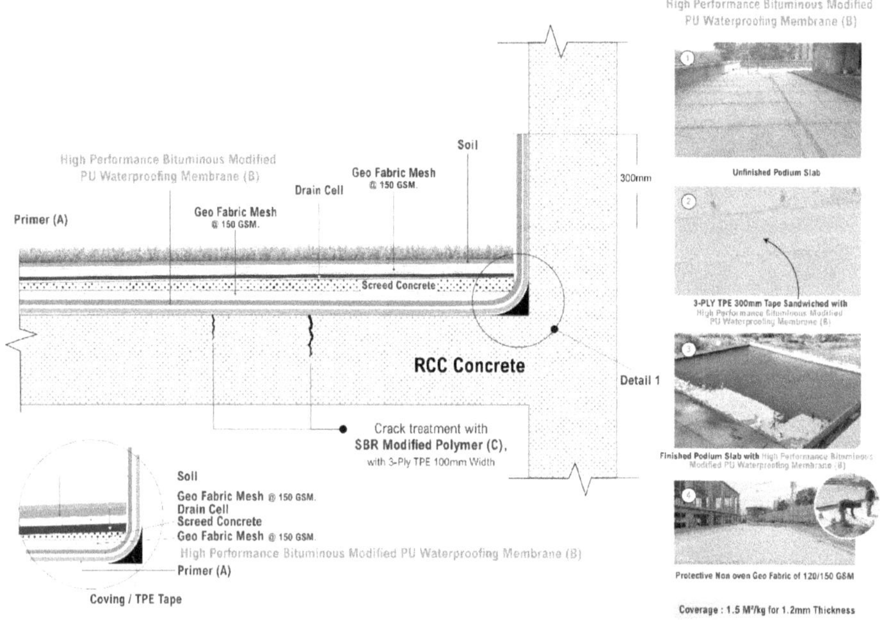

Typical coverages as per desired thickness

Rate of uses			
	Kg/m²		
	Standard	At Least	At Most
1 Coat	0.68	0.65	0.75
2 Coats	0.68	0.60	0.75
Total 2 Coats	1.35	1.20	1.50

Thickness (micron)		
	At Least	At Most
1 Coat	500	600
2 Coats	500	600
Total 2 Coats	1000	1200

Protection after Waterproofing Application:

Providing and laying a M20 Grade Screed concrete as per desired thickness for Crystalline Admixture @1% by weight of Cement, over the waterproofing coated surface.

Finishing the surface with trowel and making a thread impression of 600 mm x 600mm

Curing for 7days

Special Note:

1. Air curing of the PU will be depending on the humidity conditions at the site during the application time will be considered not on TDS doc...
2. Thickness of the film after a primer followed by 2 coats may vary depends on the concrete surface...
3. Application of PU coating to be avoided during peak monsoon season because of minimum curing period is recommended by a PU chemistry.

BOQ

S.NO	Descriptions	Unit	Qty	Rate	Amount
1.	Supply & Apply water based surface saturation primer composed of Acrylic emulsion polymer, selected special additives in water as a medium. The product has a Density of 1.01 g.cc which helps it saturate the surface it is applied upon. Apply after blending in clean water in 1:3 ratio to be sprayed using HVLP Spray Equipment. To be applied at diluted mix of 12sqm/kg. **As per Code DS-02- SatP-DryShield** Waterproofing	Sft			
2	Supply & Apply Waterproofing treatment with **Bituminous Modified PU Waterproofing Membrane (B) @ 1.6 Kg / m^2 with** a best quality of single component bituminous modified PU followed with one coat of primer and all construction joints are cut open in V groove and packed with **Structural Grade Repair Mortar** and during tacky condition applied with **3- Ply TPE Tape.** Followed by a bond coat as per the manufacturer specifications and Including surface preparation, required all chemicals, tools, labour, supervision, etc. surface preparation tools, labor, joint treatments, supervision etc.	Sft			
3.	Laying a Non oven Geotextile before screed concrete by loosely laying.				
				Total	

6. Waterproofing of Retaining Wall

Nature of Work: High Performance Bituminous Modified PU Waterproofing Membrane (B)

Primary Coating Material: Surface Saturating Acrylic Primer (A), Bituminous Modified PU Waterproofing Membrane (B)

Ancillary Materials: SBR Modified Polymer, Integral Waterproofing Liquid Admixture, Premix Crystalline Injection Grout, Structural Grade Repair Mortar, Cement & Sand

Tools & Tackles: Angle Grinder, Cutting Machine, Grinding Cup, Hammer, Chisel, Grouting Pump, Nozzle, Paddle mixer and Application Roller

METHOD STATEMENT:

SURFACE PREPARATION:

Mechanically grind the entire surface of roof to remove any black fungus material and also the loosely laid cement particles by using a grinding machine with an attachment of brush and if any cracks observed after cleaning mark those cracks and cut open the cracks by making grove so that the surface must be strong & free from dirt, dust & remove loose particles. Clean thoroughly by wire brush & Jet water wash.

CRACK TREATMENT:

Fine Cracks if any are made open by routing-out to a minimum groove and bond coat of acrylic filler is applied and is sealed with **Crack Sealer.**

Deep cracks on the slab are treated with **Low Viscous Epoxy free flow grout** by manually pouring or injection grouted in the groove.

CONSTRUCTION JOINTS:

- All construction joints are made open by cutting machine to a minimum ¾" x ¾ "groove.
- **SBR/Acrylic Modified Polymer** to be applied in the above prepared joints before with PM repair mortar.
- When bond coat is tacky it is sealed with site prepared polymer modified mortar OR with **Structural Grade Repair Mortar** workable PMM.
- All construction joints are treated with **3-PLY TPE 75 /100mm** elastomeric tape sandwiched with **Cementitious Adhesive**

COVING AT THE JUNCTION JOINTS:

- The interface joint of the floor and wall will cut will be open where ever required and clean the surface, giving one bond coat and packed with the site prepared modified mortar a specialized non – shrink cementitious compound with a ratio of 1:4 CM.
- The coving will be done as curve or taper as per the site conditions with 2" x 2" approx... With polymer modified mortar with **SBR/Acrylic Modified Polymer.**

CONSTRUCTION / STARTER JOINTS @ SLAB / PARAPET WALL JUNCTION:

Sealing the starter /construction joints at roof slab / parapet wall joints with **3-PLY TPE** 300mm elastomeric tape sandwiched with **Cementitious Adhesive**

BORE PACKING:

- Chipping around the Rainwater down take Pipes.
- All pipe prepared joints to be grouted with non-shrink cementitious flowable grout mix.

GROUTING:

- If required depending on the site condition after pond testing grouting will be done by using hand operated grout machine.
- Our cementitious grouting consists of **factory packed Cementitious Crystalline grout.**

Starter Joints @ Raft / Retaining Wall

Sealing the starter /construction joints at Starter & Retaining wall with **3-PLY TPE** 300mm elastomeric tape sandwiched with **High Performance Bituminous Modified PU Waterproofing Membrane (B)** when the structure has been casted. However, its recommended to use a **Swellable Water Bar** at the starter & retaining wall joint at the construction stage itself.

Waterproofing Application:

1. Before application of Polyurethane Liquid Membrane coating surface should be 100% dry condition or moisture free. (Moisture to be checked with moisture meter and should not be more than 5%).
2. Apply **Surface Saturating Acrylic Primer (A)** with (1:3 dilution)
3. Application of base coat to be carried out by using **High Performance Bituminous Modified PU Waterproofing Membrane (B)**
4. single component with a dilution with water (1:1) by using a roller.
5. Allow primer to dry for 5 to 8 Hours.
6. Application of First coat of **High Performance Bituminous Modified PU Waterproofing Membrane (B)**
7. @ 0.60 to 0.75kg/m²/coat High Performance Polyurethane Liquid membrane will be applied by roller on the primed surface

8. Applied First coat of **High Performance Bituminous Modified PU Waterproofing Membrane (B)** should be allowed it to air curing for about 10 – 12 hrs interval for the second coat.

9. After every coat checking / clearance given by client engineer before proceeding to next coat. After dried hours second coat to be applied in perpendicular direction to the first coat using **High Performance Bituminous Modified PU Waterproofing Membrane (B)** with 10% water dilution to be applied using roller and thickness may vary depending on the concrete surface. Consumption of the material with two coats of **High Performance Bituminous Modified PU Waterproofing Membrane (B)** will be around 1.5 Kg/M^2 to achieve 1.2mm thickness.

10. Laying of Protective **HDPE Drain Board** over the membrane

11. The system has to be left to atmospheric curing for 5-7 days before back filling of the soil.

Special Note:

- Air curing of the PU will be depending on the humidity conditions at the site during the application time will be considered not on TDS doc...

- Thickness of the film after a primer followed by 2 coats may vary depends on the concrete surface...

- Application of PU coating to be avoided during peak monsoon season because of minimum curing period is recommended by a PU chemist.

<table>
<tr><td align="center">**Recommended Waterproofing System**</td></tr>
<tr><td>

1. **Primer: Single coat Surface Saturating Acrylic Primer (A) with Bituminous Modified PU Waterproofing Membrane (B)**

2. **Application: Two Coats Application with Bituminous Modified PU Waterproofing Membrane (B)**

</td></tr>
</table>

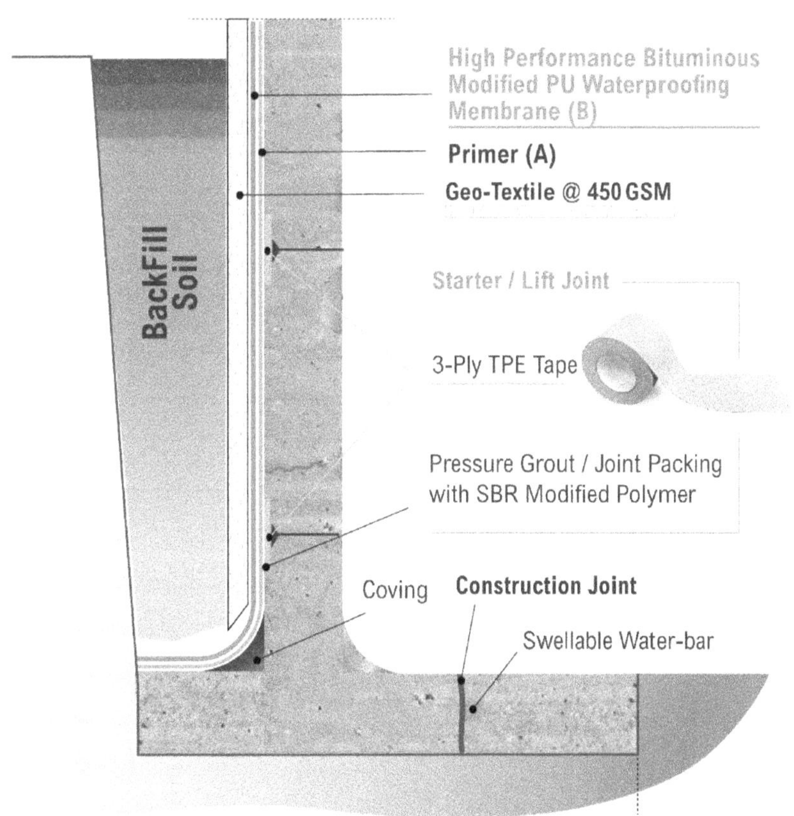

S.NO	Descriptions	Unit	Qty	Rate	Amount
1.	Supply & Apply water based surface saturation primer composed of Acrylic emulsion polymer, selected special additives in water as a medium. The product has a Density of 1.01 g.cc which helps it saturate the surface it is applied upon. Apply after blending in clean water in 1:3 ratio to be sprayed using HVLP Spray Equipment. To be applied at diluted mix of 12sqm/kg. **As per Code DS-02- SatP-DryShield** Waterproofing	Sft			
2	**Providing Waterproofing treatment with** **High Performance Bituminous Modified PU Waterproofing Membrane (B)** **@ 1.5 Kg / m^2 with a best quality of single component bitumen modified PU followed with one coat of Surface Saturating Acrylic Primer (A) and all construction joints are cut open in V groove and packed with Structural Grade Repair Mortar and during tacky condition applied with 3 - Ply TPE Tape. Followed by a bond coat as per the manufacturer specifications and Including surface preparation, required all chemicals, tools, labour, supervision, etc. surface preparation tools, labour, joint treatments, supervision etc.**	Sft			
3.	Laying a Non oven Geotextile.	Sft			
				Total	

7. Waterproofing of STP Tanks

Nature of Work: Primary Waterproofing on Concrete Surface.

Secondary Protection: Solvent Free Epoxy Coating on Protective Plaster.

Primary Coating Material: Surface Saturating Acrylic Primer (A) & Acrylic Cementitious Waterproofing Coating (B)

Ancillary Materials: SBR Modified Polymer, Integral Waterproofing Liquid Admixture, Plasticised Grout Admixture, Structural Grade Repair Mortar, Cement & Sand

Secondary Chemical Resistant Epoxy Coating Material

Tools & Tackles: Angle Grinder, Cutting Machine, Grinding Cup, Hammer, Chisel, Grouting Pump, Nozzle, Paddle mixer and Application Roller

Method Statement:

SURFACE PREPARATION:

Mechanically grind the entire surface of roof to remove any black fungus material and also the loosely laid cement particles by using a grinding machine with an attachment of brush and if any cracks observed after cleaning mark those cracks and cut open the cracks by making grove so that the surface must be strong & free from dirt, dust & remove loose particles. Clean thoroughly by wire brush & Jet water wash.

CRACK TREATMENT:

Fine Cracks if any are made open by routing-out to a minimum groove and bond coat of acrylic filler is applied and is sealed with **Crack Sealer.** Deep cracks on the slab are treated with **Low Viscous Epoxy free flow grout** by manually pouring or injection grouted in the groove.

CONSTRUCTION JOINTS:

- All construction joints are made open by cutting machine to a minimum ¾" x ¾ "groove.
- **SBR/Acrylic Modified Polymer** to be applied in the above prepared joints before with PM repair mortar.
- When bond coat is tacky it is sealed with site prepared polymer modified mortar OR with **Structural Grade Repair Mortar** workable PMM.
- All construction joints are treated with **3-PLY TPE** 75 /100mm elastomeric tape sandwiched with **Cementitious Adhesive**

COVING AT THE JUNCTION JOINTS:

- The interface joint of the floor and wall will cut will be open where ever required and clean the surface, giving one bond coat and packed with the site prepared modified mortar a specialized non – shrink cementitious compound with a ratio of 1:4 CM.
- The coving will be done as curve or taper as per the site conditions with 2" x 2" approx... With polymer modified mortar with **SBR/Acrylic Modified Polymer.**

CONSTRUCTION / STARTER JOINTS @ SLAB / PARAPET WALL JUNCTION:

Sealing the starter /construction joints at roof slab / parapet wall joints with **3-PLY TPE** 300mm elastomeric tape sandwiched with **Cementitious Adhesive**

BORE PACKING:

- Chipping around the Rainwater down take Pipes.
- All pipe prepared joints to be grouted with non-shrink cementitious flowable grout mix.

GROUTING:

- If required depending on the site condition after pond testing grouting will be done by using hand operated grout machine.
- Our cementitious grouting consists of **factory packed Cementitious Crystalline grout.**

Waterproofing Application:

Primary Application: Two Component Cementitious

- Dilute **Surface Saturating Acrylic Primer (A)** with water in proportion of 1:3 and stir well until a uniform consistency is achieved. Apply a single coat of diluted **Primer (A)** on the cleaned substrate & allow it to dry for 6 to 7 hours before commencing subsequent waterproofing coating system.
- Apply Two Component, elastomeric acrylic cementitious Fibre reinforced **Acrylic Cementitious Waterproofing Coating (B)** over the Primed surface. The Two component system Part A: two-component Cementitious Waterproofing System Powder and Part B: **Acrylic Cementitious Waterproofing Coating (B)** liquid has to be mixed in the same proportion as recommended by manufacturer. Apply using a stiff brush, roller or spray to obtain the desired thickness on horizontal and Vertical surfaces with a material consumption of approx.1.5-1.6 Kg/ m^2 for two coats with an interval of 6 -7 hrs based on the site condition.

Package and Mixing Ratio:

Primer (A) Consisting of 20 KG Package:

Liquid component 20 kg

Water component 60 kg

Acrylic Cementitious Waterproofing Coating (B) Consisting of 30 Kg Package:

Part A: Powder component 20 kg

Part B: Liquid Acrylic Polymer component 10 kg

Note: No extra water to be added.

Protection Plaster after Waterproofing Application:

Providing protective plaster in wall with C.M 1:4 of thickness 15 to 20mm on walls with **SBR/Acrylic Modified Mortar** over the waterproofing coated surface up to coated surface height. Floor has to be done SBR/Acrylic Modified screed concrete with M20 grade with proper slope

Curing time has to be carried out for minimum 3-4 days

Secondary Application: Chemical Resistant Epoxy Coating

Once the system is 100% dry a surface treatment will be carried out by using a solvent free Epoxy Coating with two coats using **Solvent Free Epoxy Coating.**

Package and Mixing Ratio:

Solvent Free Epoxy Coating - Part A: 5kg, Part B: 1 kg

Recommended Waterproofing System
• **Primer: Single coat Surface Saturating Acrylic Primer (A)** • **Application: Two Coats Application with Acrylic Cementitious Waterproofing Coating (B)** • **Protective Plaster** • **Two Coats of Solvent Free Chemical Resistant Epoxy Coating**

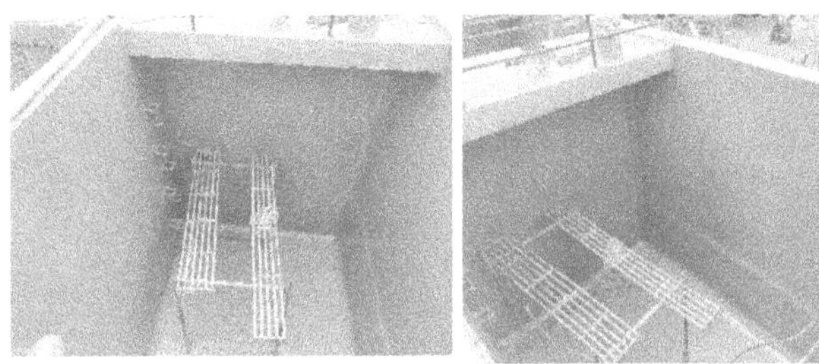

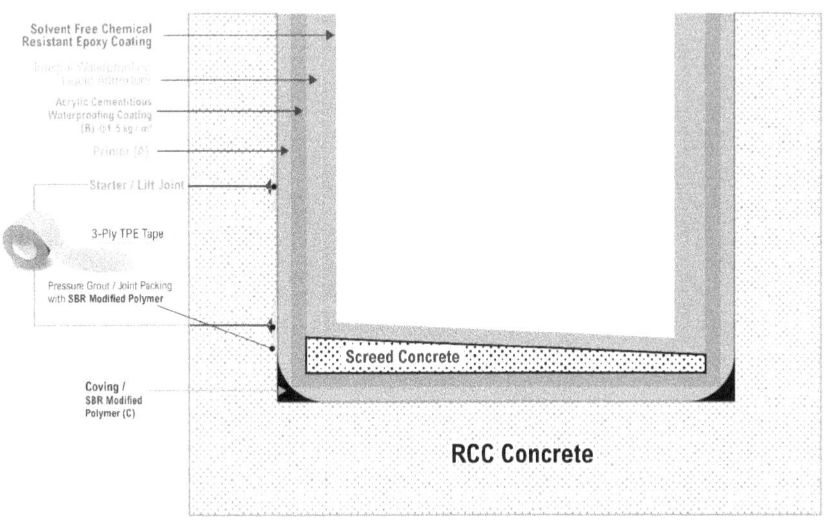

BOQ

S.NO	Descriptions	Unit	Qty.	Rate	Amount
1.	**STP Tank: Primary Waterproofing Coating** Providing Waterproofing treatment with **Acrylic Cementitious Waterproofing Coating (B) @ 1.5 Kg / m²** a best quality of two component elastomeric coating system and followed with one coat of penetrating waterproofing repellent primer for concrete as **Primer (A)** and all construction joints are cut open in V groove and packed with **Structural Grade Repair Mortar** and during tacky condition applied with **3-PLY TPE TAPE 75-100mm Width.** Followed by a bond coat **as** per the manufacturer specifications and including surface preparation, required all chemicals, tools, labour, supervision, etc... surface preparation tools, labor, joint treatments, supervision etc...	M²			
2.	**Secondary Chemical Resistant Solvent Free Epoxy Coating.** Providing protective surface treatment with **Solvent Free chemical resistant Epoxy Coating** in two coats application by using roller.	M²			
			Total Qty		

8. Specifications for Swimming Pool Waterproofing

Nature of Work: Fibre Reinforced Acrylic Cementitious Elastomeric Waterproofing.

Primary Coating Material: Surface Saturating Acrylic (A), 2 Component Acrylic Cementitious Waterproofing

Tools & Tackles: Angle Grinder, Cutting Machine, Grinding Cup, Hammer, Chisel, Grouting Pump, Nozzle, Paddle mixer and Application Roller

Method Statement:

SURFACE PREPARATION:

Mechanically grind the entire surface of roof to remove any black fungus material and also the loosely laid cement particles by using a grinding machine with an attachment of brush and if any cracks observed after cleaning mark those cracks and cut open the cracks by making grove so that the surface must be strong & free from dirt, dust & remove loose particles. Clean thoroughly by wire brush & Jet water wash.

CRACK TREATMENT:

Fine Cracks if any are made open by routing-out to a minimum groove and bond coat of acrylic filler is applied and is sealed with **Crack Sealer.** Deep cracks on the slab are treated with **Low Viscous Epoxy free flow grout** by manually pouring or injection grouted in the groove.

CONSTRUCTION JOINTS:

- All construction joints are made open by cutting machine to a minimum ¾" x ¾ "groove.
- **SBR Modified Polymer** to be applied in the above prepared joints before with PM repair mortar.
- When bond coat is tacky it is sealed with site prepared polymer modified mortar OR with **Structural Grade Repair Mortar** workable PMM.
- All construction joints are treated with **3-PLY TPE 75 /100mm** elastomeric tape sandwiched with **Cementitious Adhesive**

COVING AT THE JUNCTION JOINTS:

- The interface joint of the floor and wall will cut will be open where ever required and clean the surface, giving one bond coat and packed with the site prepared modified mortar a specialized non – shrink cementitious compound with a ratio of 1:4 CM.
- The coving will be done as curve or taper as per the site conditions with 2" x 2" approx... With polymer modified mortar with **SBR/Acrylic Modified Polymer.**

CONSTRUCTION / STARTER JOINTS @ SLAB / PARAPET WALL JUNCTION:

Sealing the starter /construction joints at roof slab / parapet wall joints with **3-PLY TPE** 300mm elastomeric tape sandwiched with **Cementitious Adhesive**

BORE PACKING:

- Chipping around the Rainwater down take Pipes.
- All pipe prepared joints to be grouted with non-shrink cementitious flowable grout mix.

GROUTING:

- If required depending on the site condition after pond testing grouting will be done by using hand operated grout machine.
- Our cementitious grouting consists of **factory packed Cementitious Crystalline grout.**

Waterproofing Application:

Primary Application: Two Component Cementitious

- Dilute **Surface Saturating Acrylic Primer (A)** with water in proportion of 1:3 and stir well until a uniform consistency is achieved. Apply a single coat of diluted **Primer (A)** on the cleaned substrate & allow it to dry for 6 to 7 hours before commencing subsequent waterproofing coating system.
- Apply Two Component, elastomeric acrylic cementitious Fibre reinforced **Acrylic Cementitious Waterproofing Coating (B)** over the Primed surface. The Two component system Part A: two-component Cementitious Waterproofing System Powder and Part B: **Acrylic Cementitious Waterproofing Coating (B)** liquid has to be mixed in the same proportion as recommended by manufacturer. Apply using a stiff brush, roller or spray to obtain the desired thickness on horizontal and Vertical surfaces with a material consumption of approx.1.5-1.6 Kg/m² for two coats with an interval of 6 -7 hrs based on the site condition.

Package and Mixing Ratio:

Primer (A) Consisting of 20 KG Package:

Liquid component 20 kg

Water component 60 kg

Acrylic Cementitious Waterproofing Coating (B) Consisting of 30 Kg Package:

Part A: Powder component 20 kg

Part B: Liquid Acrylic Polymer component 10 kg

Note: No extra water to be added.

Protection Plaster after Waterproofing Application:

Providing protective plaster in wall with C.M 1:4 of thickness 35-40 mm on walls with **SBR/Acrylic Modified Mortar** over the waterproofing coated surface up to coated surface height. Floor has to be done SBR/Acrylic Modified screed concrete with M20 grade with proper slope

Secondary Application:

- Dilute **Surface Saturating Acrylic Primer (A)** with water in proportion of 1:3 and stir well until a uniform consistency is achieved. Apply a single coat of diluted **Primer (A)** on the cleaned substrate & allow it to dry for 6 to 7 hours before commencing subsequent waterproofing coating system.

Curing time has to be carried out for minimum 3-4 days

Recommended Waterproofing System
1. Crystalline Admixture in Concrete 2. Primer: Single coat Surface Saturating Acrylic Primer 3. Application: Two Coats Application with high elongation 2 component Acrylic Cementitious 4. Protective Plaster 5. Adhesive and Joint Filler for Tiles

BOQ

S.NO	Descriptions	Unit	Qty.	Rate	Amount
1	Supply & Apply a water based surface saturation primer composed of Acrylic emulsion polymer, selected special additives in water as a medium. The product has a Density of 1.01 g.cc which helps it saturate the surface it is applied upon. Apply after blending in clean water in 1:3 ratio to be sprayed using HVLP Spray Equipment. To be applied at diluted mix of 12sqm/kg. Apply Corner Tapes on all joints using Cementous adhesive, **As per Code DS-02- SatP-DryShield** Waterproofing				
2	Supply & Apply, a two component elastomeric coating waterproofing system based on cementitious binders reinforced with micro fibres, special additives, fine-grained selected aggregates, high grade acrylic polymers to provide superior water protection. The Powder: Polymer ratio 1:1, Elongation >200%, Tensile Strength (N/mm2) 0.82, Water Absorption ASTM D 570 1%, Static Crack bridging ability (mm) >2, Shore A Hardness 70.. **The coating should be applied on the pre saturated surface @ 1.4kg./sqm to achieve a DFT of 1mm. As per Dryshield Code DS/2KCemt/DS 03 DryShield Waterproofing**				
3	Supply & Apply a water based surface saturation primer composed of Acrylic emulsion polymer, selected special additives in water as a medium. The product has a Density of 1.01 g.cc which helps it saturate the surface it is applied upon. Apply after blending in clean water in 1:3 ratio to be sprayed using HVLP Spray Equipment. To be applied at diluted mix of 12sqm/kg. **As per Code DS-02- SatP-DryShield** Waterproofing	M²			
			Total Qty		
			IGST 18%		
			Grand Total		

Thank You

I would like to express my gratitude to Mr. Samir Surlekar, Mr. J. Srinivas Prasad, Mr. Sunny Surlekar, and Mr. Udit Narayan Dass for their unwavering support and guidance over the past 25 years. Their expertise in Waterproofing and Building Envelope has been invaluable in helping me understand the nuances of this field. They answered my queries and motivated me to learn, experiment, and innovate. Their support has been consistent and continues to this day.